DISCRETE MATHEMATICS
FOR COMPUTING

MATHEMATICS AND ITS APPLICATIONS

Series Editor: G. M. BELL, Professor of Mathematics, King's College London, University of London

STATISTICS, OPERATIONAL RESEARCH AND COMPUTATIONAL MATHEMATICS

Editor: B. W. CONOLLY, Emeritus Professor of Mathematics (Operational Research), Queen Mary College, University of London

Mathematics and its applications are now awe-inspiring in their scope, variety and depth. Not only is there rapid growth in pure mathematics and its applications to the traditional fields of the physical sciences, engineering and statistics, but new fields of application are emerging in biology, ecology and social organization. The user of mathematics must assimilate subtle new techniques and also learn to handle the great power of the computer efficiently and economically.

The need for clear, concise and authoritative texts is thus greater than ever and our series endeavours to supply this need. It aims to be comprehensive and yet flexible. Works surveying recent research will introduce new areas and up-to-date mathematical methods. Undergraduate texts on established topics will stimulate student interest by including applications relevant at the present day. The series will also include selected volumes of lecture notes which will enable certain important topics to be presented earlier than would otherwise be possible.

In all these ways it is hoped to render a valuable service to those who learn, teach, develop and use mathematics.

Mathematics and its Applications

Series Editor: G. M. BELL, Professor of Mathematics, King's College London, University of London

Anderson, I.	Combinatorial Designs: Construction Methods
Artmann, B.	Concept of Number: From Quaternions to Monads and Topological Fields
Arczewski, K. & Pietrucha, J.	Mathematical Modelling in Discrete Mechanical Systems
Arczewski, K. and Pietrucha, J.	Mathematical Modelling in Continuous Mechanical Systems
Bainov, D.D. & Konstantinov, M.	The Averaging Method and its Applications
Baker, A.C. & Porteous, H.L.	Linear Algebra and Differential Equations
Balcerzyk, S. & Jösefiak, T.	Commutative Rings
Balcerzyk, S. & Jösefiak, T.	Commutative Noetherian and Krull Rings
Baldock, G.R. & Bridgeman, T.	Mathematical Theory of Wave Motion
Ball, M.A.	Mathematics in the Social and Life Sciences: Theories, Models and Methods
de Barra, G.	Measure Theory and Integration
Bartak, J., Herrmann, L., Lovicar, V. & Vejvoda, D.	Partial Differential Equations of Evolution
Bell, G.M. and Lavis, D.A.	Statistical Mechanics of Lattice Models, Vols. 1 & 2
Berry, J.S., Burghes, D.N., Huntley, I.D., James, D.J.G. & Moscardini, A.O.	Mathematical Modelling Courses
Berry, J.S., Burghes, D.N., Huntley, I.D., James, D.J.G. & Moscardini, A.O.	Mathematical Modelling Methodology, Models and Micros
Berry, J.S., Burghes, D.N., Huntley, I.D., James, D.J.G. & Moscardini, A.O.	Teaching and Applying Mathematical Modelling
Blum, W.	Applications and Modelling in Learning and Teaching Mathematics
Brown, R.	Topology: A Geometric Account of General Topology, Homotopy Types and the Fundamental Groupoid
Burghes, D.N. & Borrie, M.	Modelling with Differential Equations
Burghes, D.N. & Downs, A.M.	Modern Introduction to Classical Mechanics and Control
Burghes, D.N. & Graham, A.	Introduction to Control Theory, including Optimal Control
Burghes, D.N., Huntley, I. & McDonald, J.	Applying Mathematics
Burghes, D.N. & Wood, A.D.	Mathematical Models in the Social, Management and Life Sciences
Butkovskiy, A.G.	Green's Functions and Transfer Functions Handbook
Cartwright, M.	Fourier Methods: for Mathematicians, Scientists and Engineers
Cerny, I.	Complex Domain Analysis
Chorlton, F.	Textbook of Dynamics, 2nd Edition
Chorlton, F.	Vector and Tensor Methods
Cohen, D.E.	Computability and Logic
Cordier, J.-M. & Porter, T.	Shape Theory: Categorical Methods of Approximation
Crapper, G.D.	Introduction to Water Waves
Cross, M. & Moscardini, A.O.	Learning the Art of Mathematical Modelling
Cullen, M.R.	Linear Models in Biology
Dunning-Davies, J.	Mathematical Methods for Mathematicians, Physical Scientists and Engineers
Eason, G., Coles, C.W. & Gettinby, G.	Mathematics and Statistics for the Biosciences
El Jai, A. & Pritchard, A.J.	Sensors and Controls in the Analysis of Distributed Systems
Exton, H.	Multiple Hypergeometric Functions and Applications
Exton, H.	Handbook of Hypergeometric Integrals
Exton, H.	q-Hypergeometric Functions and Applications
Faux, I.D. & Pratt, M.J.	Computational Geometry for Design and Manufacture
Firby, P.A. & Gardiner, C.F.	Surface Topology
Gardiner, C.F.	Modern Algebra

Series continued at back of book

DISCRETE MATHEMATICS FOR COMPUTING

A. VINCE, B.Sc., M.Sc., M.A., F.I.M.A.
Department of Computer Studies and Mathematics, Bristol Polytechnic

and

C. MORRIS, B.A.(M.A.), M.Sc., Ph.D., L.T.C.L.
School of Industrial and Business Studies, University of Warwick

ELLIS HORWOOD
NEW YORK LONDON TORONTO SYDNEY TOKYO SINGAPORE

First published in 1990 by
ELLIS HORWOOD LIMITED
Market Cross House, Cooper Street,
Chichester, West Sussex, PO19 1EB, England
A division of
Simon & Schuster International Group

© Ellis Horwood Limited, 1990

All rights reserved. No part of this publication may be
reproduced, stored in a retrieval system, or transmitted,
in any form, or by any means, electronic, mechanical,
photocopying, recording or otherwise, without the prior
permission, in writing, of the publisher

Typeset in Times by Ellis Horwood Limited
Printed and bound in Great Britain
by Hartnolls, Bodmin

British Library Cataloguing in Publication Data

Vince, A.
Discrete mathematics for computing.
1. Mathematics. Discrete functions.
I. Title   II. Morris, C.
511.33
ISBN 0–13–217522–3 (Library Edn.)
ISBN 0–13–217514–2 (Student Edn.)

Library of Congress Cataloging-in-Publication Data available

Table of contents

Preface 7
   1. **Introduction to discrete mathematics** 9
   2. **Sets and set algebra** 25
   3. **Elementary logic** 47
   4. **Data representation and manipulation** 70
   5. **Graph theory** 96
   6. **Digraphs** 124
   7. **Trees** 152
   8. **Relations and functions** 194
   9. **Counting and probability** 219
  10. **Error-correcting codes** 241
Suggestions for further reading 264
Answers to exercises and problems 265
Index 295

Preface

The decision to write this book was taken chiefly as a result of our experience in designing syllbuses for, and ultimately teaching, mathematics units on HND and HNC courses in Computer Studies, and on the first year of a BA in Systems Analysis. While it is generally agreed that such students certainly need some mathematical input, it was more difficult to determine precisely what the nature of that input should be; and even when we had, in consultation with our computing and systems colleagues, arrived at a core of topics which needed to be covered, we were hampered by the lack of a suitable text.

There are numerous excellent books, mostly by American authors, with titles which are variations on 'Discrete math for computer science', but it seemed to us that few of these were suited, in terms of language, style and approach, to the needs of British students, especially at the sub-degree level. The mathematics was often presented in quite a formal way, and complex computing applications were presented which required a good deal of knowledge on the part of the reader.

Accordingly, we have set out to write a text which in both language and structure will be 'user-friendly' to students who approach the subject with little formal mathematical background. While we hope that we have not fudged any mathematical issues, we make no claim to rigour; instead we aim to give the reader a sound grasp of some important basic ideas, and to illustrate those ideas by reference to relevant, yet simple, applications.

There is, as already mentioned, no consensus as yet on the topics which need to be covered in a course of this type, though most would agree that set theory, some combinatorial mathematics, and the fundamentals of Boolean algebra form an irreducible minimum. Our selection of topics is therefore to some extent personal; the reader will, for example, find only a brief mention of finite state machines, but will discover quite substantial chapters on undirected and directed graphs.

One choice which immediately confronts the authors of a book like this is whether or not it should include examples of algorithms written in a specific language. We have decided, in the interests of generality, not to take this route, but instead to make use of 'action diagrams' to present algorithms in a structured, yet language-free manner. This approach, introduced by Martin and McClure in *Action diagrams: clearly structured program design* (Prentice Hall, 1985) seems to us to enable students to get to grips with the essential structure of an algorithm without becoming too worried as to whether a program would actually run.

We are glad to have this opportunity of thanking all our past students, particularly those of the BA Systems Analysis at Bristol Polytechnic, for the stimulation which teaching them has given us, and for giving rise, by their comments on examples and their discussion in seminars, to many of the ideas embodied in this book. It is also a pleasure to thank Peter Shiarly of the University of Bristol and Johanna Vince for drawing many of the diagrams.

Finally, we hope that any readers who come across errors or obscurities in the book, or who would like to comment on it in any way, will let us know. Writing a book is a learning process which we hope will not finish with its publication. We look forward to hearing from you!

AJV
CM

February 1989

1
Introduction to discrete mathematics

This book is designed to introduce you to a type of mathematics, rather different from the kind you probably studied at school, which has wide-ranging applications to the solution of problems in computing and data processing. Since the ideas will initially be somewhat unfamiliar, and even the name 'discrete mathematics' may not mean very much to you at present, we begin this chapter with a discussion of just what 'discrete mathematics' involves.

Because we are primarily interested in the application of our mathematical ideas to computing, it is important to be able to give instructions for mathematical processes in a structured and systematic fashion. Such structured sets of instructions are what is meant by the term *algorithms*, and we devote the remainder of this chapter to introducing a particularly neat and simple way of specifying algorithms, called the *action diagram*. We will use these diagrams throughout the rest of the book whenever an algorithm occurs, so you need to become thoroughly familiar with them.

The remaining chapters of the book do not necessarily have to be read in the order in which they occur. In fact, the topics we cover are so interconnected that we sometimes felt some kind of loose-leaf structure for the text, in which sections could be re-ordered at will, would be more suitable than a conventional front-to-back book! We have selected what we feel to be the most logical order, but some sections can be omitted without affecting what comes later. This chapter is essential to everything which follows; Chapters 2, 3, and 4 are also fundamental, and need to be read as a sequence; Chapters 5, 6, and 7 form a group, but if you wish they may be omitted, as Chapters 9 and 10 make little reference to them.

A word about the exercises: you will find that some of these are embedded in the text, while others are gathered together as problems at the end of chapters. You should if at all possible work the text examples as you come to them before proceeding with your reading; they will help you to become familiar with the ideas. It is all too easy to read a mathematical text, imagine that you understand what is going on, and then find when you reach the end-of-chapter exercises that things are not quite so clear as you had thought!

Finally, we are very conscious that an introductory book of this kind can only give you a glimpse of what is a vast and ever-expanding subject. We have therefore provided some 'Suggestions for further reading' on p. 264; if you become particularly interested in any of the topics we cover, you will find there a list of books which will enable you to examine the various areas in more depth.

1.1 WHY DISCRETE MATHS?

The adjective 'discrete' is defined in the Oxford English Dictionary as 'Separate, detached from others, individually distinct. Opposed to *continuous*'. As beginners in maths, at around the age of four or five you were using discrete maths — 'one and two makes three' is an example. You applied this arithmetic to distinct objects like apples, books or pencils, and the mathematics you used — addition, multiplication, subtraction and division–reflected this.

Discrete maths is appropriate when objects are counted (rather than weighed or measured), and when we consider relations between one set of objects and another set. We will find that processes in discrete maths generally terminate after a finite number of steps, and that the terms 'infinite' and 'infinitesimal' which occur so frequently in the differential calculus are unlikely to occur. It's worth reminding ourselves that a digital computer is a *finite state machine*. That is to say, just as a simple object like a light switch has only two possible states — switched to 'on' or 'off' — so the world's largest computer has a finite number of possible states at any point in time, though that number may be unimaginably large. A computer is also finite in the sense that it cannot always produce exact results. For instance a computer cannot store the result of the division sum 1/7 in exact form, any more than we can write down the answer as an exact decimal.

1.2 THE IDEA OF MODELLING WITH MATHEMATICS

One of the reasons why aeroplanes fly, communications satellites remain stationary over the Earth, and the Severn Bridge is still standing, apart from the experience and skills of the designers and makers is that there are consistent laws of physics, and that these laws can be expressed as mathematics. A whole science of mathematical modelling exists to evaluate and assist in the design of aeroplanes, satellites, bridges and hundreds of other artefacts. But notice that it's vital to apply appropriate mathematics, otherwise we get wrong or nonsensical answers. Take the case at the start of the chapter, where we said that one and two makes three. If we add one litre of water at temperature of 15°C to two litres of water at 25°C, then we get approximately three litres of water. Notice that we have assumed the word 'add' is equivalent to the mathematical operation '+' (and is also equivalent to pouring water from one container into another). Addition is valid since, in this case, mixing amounts to the same thing as addition of volumes. However, in the case of the *temperatures* it would be a mistake to apply simple addition — temperatures don't combine in a simple additive way, and the resultant temperature would not be 40°C. So in general when we apply mathematics, we must be careful that it is mathematics appropriate to the problem.

1.3 AN ALGEBRAIC MODEL

We will look at an example which is familiar to students of school algebra, leaving aside for the moment the question whether or not this is an example of discrete mathematics (as it's the modelling that's of interest at present). The example concerns the buying habits of students.

Suppose that I noticed two students passing successively through the checkout in the college refectory. The first one purchased a burger and two bottles of Coke, for which he paid a total of 170 p: the other purchased three burgers and a Coke and paid 210 p.

From this information we can make a mathematical model of the situation. From the model we can find out the price of a single Coke, and the price of a single burger (don't worry that any sensible person would simply consult the price list; what we are really doing is examining the modelling process).

We can say let the price of a burger in pence be denoted by x, and similarly let the price of a Coke in pence be denoted by y (we don't say let a burger be x, and a Coke be y, as we are talking not about the items themselves, but about related prices).

Now we can make two simple mathematical statements about the two students' purchases. In each case we represent the known sums paid in terms of the supposed unit price of a burger and of a Coke respectively. The model of the situation is in the form of two equations, which employ addition, multiplication and equality (+, * and = respectively).

(1) $1*x+2*y=170=$ the bill for the first student

and

(2) $3*x+1*y=210=$ the bill for the second student

Notice how many assumptions we've made in order to set up these equations. First we assume that all burgers are the same — or at least that all burgers are sold at the same unit price. (In reality different kinds of burgers would probably cost different prices.) A second assumption is that there is no sales policy of the type 'buy two burgers — get one for half price', which might explain the large purchase by the second student, but which would oblige us to say that in some circumstances, $2+1=2.5$. We also assume tacitly that sums of money are additive — something which fortunately everyone else agrees about. Finally, in any modelling process we deliberately *ignore* a lot of information which might be relevant for other purposes — for example, we don't consider the name, age or sex of the student; and we aren't interested in the length of the queue or in last year's prices, as none of these will help us to get the answer we are looking for. An algebraic solution to the model is shown below, though there's no necessity for you to work through it.

(Each time we modify an equation we will give it a new number, as shown down the left-hand side. The information on the right-hand side explains how the new equation was obtained — check that equation (3) below *is* in fact the result of multiplying equation (2) by 2.)

We apply simple rules of algebra to these equations and get

(1) $1*x+2*y=170=$ the bill for the first student
(2) $3*x+1*y=210=$ the bill for the second student

(3) $6*x+2*y=420$ (2)*2

Then we can subtract equation (1) from equation (3) to get a new equation, which we will use as a replacement for (3):

(4) $5*x=250$ (3)−(1)

This gives us a value for x

(5) $1*x=50$ (4)/5

Now we can either substitute the value 50 for x in (1) and solve (1) or, as shown below subtract (5) from (1), to obtain

(6) $2*y=120$ (1)−(5)

This gives the final equation

(7) $1*y=60$ (6)/2

Now we have to remember that these mathematical operations were carried out on a model. The result is that x has a value of 50 and y has a value of 60, as shown by equations (5) and (7). If we recall the original meaning that we gave to x and y we can say that the answer to the original problem is that the price of a burger is 60p, and the price of a coke is 50p.

Notice that the answer can be checked by substituting the values of x and y in the original model — equations (1) and (2) — and also that our model has a modest predictive power for calculating the bill for anyone who is unimaginative enough to buy burgers and Cokes and nothing else.

This example has several discrete characteristics — we have distinct objects like students, bottles of Coke, burgers and pence, and the problem terminates in a finite number of steps — but it's the *modelling* aspect of the problem which is of interest, rather than the discrete nature (or otherwise) of the mathematics.

Clearly there are many equations which can be solved by simple algebra, but which represent continuous rather than discrete problems. For example, $4.54*x=y$ gives the number of litres y which are equivalent to x imperial gallons. Here x and y are continuous quantities — that is, they can take absolutely any value (unlike the hamburgers of the previous example, which are generally bought in integer or whole-number quantities). However, if we put $y = 3.26$ into this equation, we can solve it for x in a single step, so the discrete characteristic here is that the equation can be solved by a finite process. Of course, the solution is only an approximation, since the number of litres to a gallon is not precisely 4.54.

1.4 A MODEL IN THE FORM OF A GRAPH

Now we will look at a problem which is a genuine example of discrete maths. Suppose we are given a list of towns; a route by road between each pair of towns has been

Sec. 1.4] **A model in the form of a graph**

chosen, and its length in miles measured. We want to find a set of these routes with two properties: firstly that it should connect all towns, either directly, or indirectly via other towns; and secondly that the total mileage of the chosen routes should be a *minimum*. The table below shows the mileages. Notice that this table is a model in its own right, having been at some time derived from a map, which was in turn modelled from information provided by surveyors, road builders, geographers and so forth — the table has a long and complex pedigree!

Penzance				
281	London			
228	167	Cardiff		
542	378	380	Edinburgh	
394	268	261	308	Holyhead

In order that we can identify the set of routes with the two required properties — that is, the property of connecting all towns and the property of using a minimum total length of road — we are going to make a model of the table which has the form of a *graph* (though not the kind of graph you have been accustomed to plotting on graph paper). This graph is shown in Fig. 1.1 — and as you can see, it carries all of the

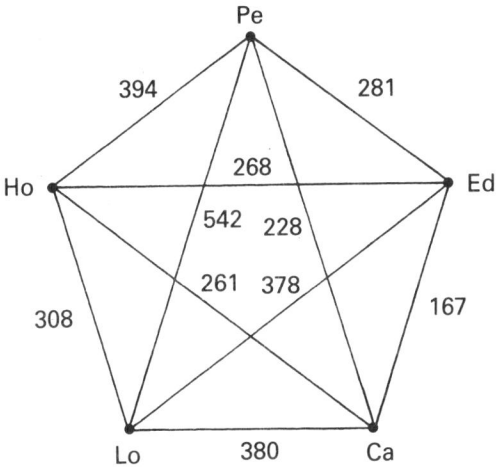

Fig. 1.1.

information from the table. The towns are represented by *vertices* (singular *vertex*), with appropriate abbreviations of names; the routes are represented by *edges*, and each edge has an associated number (or *weight*) which is equal to the mileage for the given route. This example is an example of discrete maths, not because of the mileages, which are given to the nearest integer value but are really continuous quantities, but because of the discrete nature of the objects we are dealing with:

towns and the routes between them. The terms 'vertex', 'edge', and 'weight' are frequently used terms, as you will see in Chapters 5, 6 and 7. But even at this early stage, we will see that this model of the information enables us to apply a set of rules of procedure — otherwise known as an *algorithm* — to find the best set of edges on the graph, and hence the best set of routes (defining the 'best' set of routes as the set with the minimum total mileage). You will find two algorithms for doing this in Section 7.4. Here we give a brief description of one of them, known as Kruskall's algorithm, which is fairly simple to describe and to apply (you might feel like adding the word 'obvious', but to prove that the algorithm *always* works requires a certain amount of mathematics, which you will find in Section 7.5).

One version of Kruskall's algorithm is given below. It tells us to start with a set of no edges at all, and to take edges from the graph and to put them into the set, according simple rules.

(0) Start with no edges and no vertices in the set.
(1) Look at the edges which have not so far been chosen for the set and choose the edge with lowest weight (along with the edge's two vertices); this choice is subject to rule 2 below.
(2) An edge is not eligible for choice if, with the edges already chosen for the set, it would complete a triangle, rectangle, pentagon etc. — in other words any edge which would complete a closed *circuit* may not be chosen.
(3) If the edges in the set do not yet contain all of the *vertices* of the original graph then go back to step 1.

Otherwise
(4) The set of edges corresponds to a set of routes between all the towns, and the total mileage of these routes is a minimum.

Incidentally rule 2 disallows the possibility of a circular tour (can you see why an answer containing a circular tour can always be improved on?)

The structure of edges resulting from the application of this algorithm is called a *tree*. The collection of edges which makes up the tree is shown in Fig. 1.2. The edges were selected in ascending order of edge weight (or length of route). Notice that the edge which represents the route between Holyhead and London was not selected as it would have completed a triangle.

1.5 DISCRETE DOES NOT ALWAYS MEAN SMALL

Though this simple algorithm gives the correct answer to the problem which we posed, it does *not* give the answer to a similar question which might be asked of the same data. This alternative problem will give you some feel for the idea that discrete or finite quantities, can also be large quantities (in the manner of the finite state computer we mentioned earlier). The problem is to devise a tour which starts in (say) London, visits each of the other towns once only, and then returns to London. We want to know which of all the possible tours has the minimum total mileage. The only known method which guarantees that we have the minimum mileage is to check all

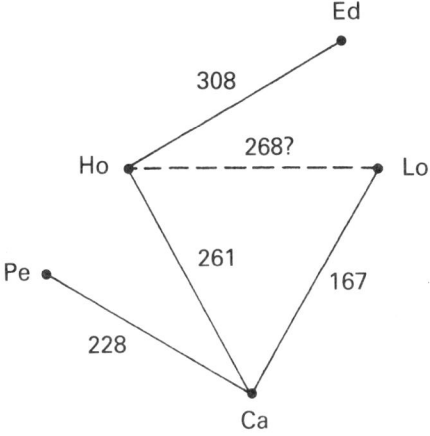

Fig. 1.2.

possible routes. It's not particularly difficult to do this, and at first sight, it looks as though it wouldn't take long.

In our five town example there are four choices on leaving London, after which we have three choices on leaving the next town, followed by two choices, and finally just the one 'choice' to get back to London. This gives us a total number of possible tours equal to 4×3×2×1, or half that number if we note that there are two ways round a circle, and that the distance is the same in both directions. (Don't worry if you don't see why we multiply the numbers; the topic of counting is dealt with in Chapter 9.) Twelve sets of easy calculations doesn't involve that much work, and we might think that 'count them all and pick the shortest' is a reasonable way to do it, especially with a computer.

However, if we applied this rule to a graph with as few as 10 vertices we would have to examine 181 440 routes (9×8×7×...×2×1)/2. And you might like to verify that an 11-vertex graph would require 1 814 400 (more than one and a half million) routes to be checked. At the time of writing, there is no known method which will guarantee that we have the best solution to this problem — apart from checking all possibilities.

1.6 ALGORITHMS AND ACTION DIAGRAMS

Much of the work we do in mathematics is of the algorithmic kind. By algorithmic, we mean that we carry out a set of *instructions* in order to get some kind of result, just as we did with Kruskall's algorithm above. In the past, you have probably used algorithmic methods, if not algorithms themselves. For example any geometric construction which you did — such a the bisection of an angle, or the construction of a right-angled triangle — could be thought of as an algorithmic proces. Similarly, if you've ever carried out a long multiplication in arithmetic or added two numbers together, you have applied some kind of algorithm.

We won't claim that the set of arithmetical rules applied to the solution of our simultaneous equations (involving the prices of burgers and Cokes) was a good algorithm if indeed it was an algorithm at all — there was far too much choice within the rules. We could have applied valid rules over and over again and could have produced hundreds of true equations, none of which gave the answer we were looking for. On the other hand, the method for finding a minimum weight tree probably *can* be thought of as an algorithm — even though some of the terms employed were not rigorously defined. For example in Kruskall's algorithm, we didn't give a mathematical rule for identifying a triangle; rather we assumed that the user would spot a potential triangle.

Now we are going to describe a useful general method for specifying an algorithm, that is, the *action diagram*. In the past you may have used a *flow chart* to describe processes or algorithms — or possibly you have met algorithms described in *pseudo-code*. In either case the intention is to give a representation of a process as an intermediate stage between a description in English and an actual program. Flow charts were fashionable up to the middle 1970s and are still used by a few people, but are generally considered to be too unwieldy to represent all but the simplest algorithms; they are also difficult to amend (a latter-day name for flow chart is 'spaghetti-diagram'). Pseudo-code is a form of program-language which looks rather like a computer program, but which is written in a limited form of English. An action diagram can be thought of as a rather more structured form of pseudo-code. We are going to concentrate on the action diagram, rather than on pseudo-code, because the action diagram is more flexible, can be applied to small algorithms (as it will be here) and can also be used to describe large commercial processes. A second feature of the action diagram is that it can be used to describe *data structures* (a simple example will be shown in this chapter, and there is a further example in Chapter 7). Again this form of diagramming has wide-ranging commercial applications. So a good reason for using action diagrams is that they have uses beyond the limited scope of this book.

An action diagram consists of just three *constructs* or building blocks which are named respectively *sequence, selection* and *repetition* — and we don't need to consider either mathematics or computing to enable us to understand the three! First we'll consider *sequence*.

Sequence
Suppose that someone travels from home to college by bicycle. The cycling part of the day consists of three actions, which are carried out every working day in the same order. These actions in order are

      Entry: cyclist at home

$$\left[\begin{array}{l} \text{Take bicycle out of garage} \\ \text{Put on hard hat} \\ \text{Pedal to college} \end{array}\right.$$

      Exit: cyclist at college

So the cyclist carries out a *sequence* of actions, and these actions are enclosed in a

Sec. 1.6] **Algorithms and action diagrams** 17

sequence bracket, as shown. To describe the state of affairs at the start of the instructions (Entry) and at the end (Exit), we also include an 'Entry' statement at the beginning, and an 'Exit' statement at the end, though these are not always included.

Example
Suppose that a student is taking a short test with five questions. If we assume that the questions must be answered in the order in which they are asked (as in an oral test, for instance) construct an action diagram to describe the process.

Answer
One possibility is

    Entry: Set of questions

$\left[\begin{array}{l}\text{answer question 1}\\\text{answer question 2}\\\text{answer question 3}\\\text{answer question 4}\\\text{answer question 5}\end{array}\right.$

    Exit: Set of answers

The student carries out a *sequence* of actions; and as before, we indicate this with the sequence bracket. The Entry and Exit conditions are rather arbitrary in this case. It's worth taking note of the fact that the structure of the test itself is also a sequence, which can be represented in action diagram form, — that is

$\left[\begin{array}{l}\text{question 1}\\\text{question 2}\\\text{question 3}\\\text{question 4}\\\text{question 5}\end{array}\right.$

    The two action diagrams which we have just seen are very short of detail — how does the cyclist get his bicycle out of the garage, or (more to the point) how does the student answer question 3? We need to remember that any description has its own particular level of detail, and that we are taking a *top-down* approach. We can always go down to whatever level of detail we need — we can give instructions to the cyclist about which side of the road to travel on, or tell the student how to construct sentences, and so forth. Any action may be decomposed into constituent actions, and the same is true of action diagrams.

Selection
Suppose we want to make sure that our cyclist doesn't ride in the rain without putting on a waterproof cape. The sequence doesn't help as it stands. Let's assume that both the hard hat and cape are stored on the bicycle. We want to say that if it's raining then the cyclist should put on his cape. This is how *selection* arises: an action is selected according to some criterion. In this case the choice (or selection) is to put on a cape if

it is raining, or by default, to do nothing. The criterion for making the choice is the state of the weather.

$$\begin{bmatrix} \text{Take bicycle out of garage} \\ \text{Put on hard hat} \\ \begin{bmatrix} \text{IF it is raining} \\ \quad \text{Put on cape} \\ \text{ENDIF} \end{bmatrix} \\ \text{Pedal to college} \end{bmatrix}$$

What we have done is to put an additional instruction into the sequence, but the new instruction is only carried out IF a particular condition is true — i.e. if it is raining. Notice that we use another sequence bracket to enclose this instruction (in another case there might be more than one instruction between IF and ENDIF).

Example
In the case of rain an improved procedure for our cyclist is that he should wear a souwester over his hard hat, and that in all circumstances he should check his brakes before he sets out. Introduce the new requirements into the action diagram.

Answer

$$\begin{bmatrix} \text{Take bicycle out of garage} \\ \text{Check the brakes} \\ \text{Put on hard hat} \\ \begin{bmatrix} \text{IF it is raining} \\ \quad \text{Put on cape} \\ \quad \text{Put on souwester} \\ \text{ENDIF} \end{bmatrix} \\ \text{Pedal to college} \end{bmatrix}$$

Note that the checking of brakes could have been carried out in a different part of the sequence, and also that we don't give any instructions about what happens if the brakes are found to be faulty. We could in fact decompose the instruction 'Check the brakes' into another sequence within the main sequence.

The idea of selection as we've described it is rather restricted, in that we either select a single action or we do nothing. This 'take it or leave it' form of selection can be extended to something more useful in two ways. The first is shown below: We say IF condition 1 then action X, IF condition 2 then action Y, IF etc., etc., for as many conditions and corresponding actions as there are. The action diagram is of the form

$$\begin{bmatrix} \text{IF condition } X \\ \quad \text{Action 1} \\ \text{IF condition } Y \\ \quad \text{Action 2} \end{bmatrix}$$

```
   ┌IF condition Z
   │ Action 3
   └ENDIF
```

and the bracket enclosing this type of sequence is known as a *selection bracket*.

Example
Suppose an exam mark is to be converted to a grade according to the following rules: 0 to 39 is a fail, 40 to 64 is a pass, 65 to 100 is a distinction. Draw the corresponding action diagram.

Answer
One possibility is

```
Entry: an exam mark
  ┌IF (0≤mark≤39)
  │  result = 'fail'
  ├IF (40≤mark≤64)
  │  result = 'pass'
  ├IF (65≤mark≤100)
  │  result = 'distinction'
  └ENDIF
Exit: a grade
```

(There is a less formal and shorter alternative version of this type of selection in Section 8.6, illustrating the algorithm called Zeller's rule.)

The second extension of this IF (or selection) construct is the ELSE construct. Suppose our cyclist doesn't like the idea of wearing a souwester over his hard hat, and decides that he will only wear his hard hat if it's not raining. The modified diagram could make use of the ELSE as follows

```
⎡ Take bicycle out of garage
⎢ Check the brakes
⎢  ┌IF it is raining
⎢  │  Put on cape
⎢  │  Put on souwester
⎢  ├ELSE
⎢  │  Put on hard hat
⎢  └ENDIF
⎢ Pedal to college
⎣
```

Example 1
The action diagram which describe the conversion of a mark to a grade would not give a result for an invalid mark such as 970. Amend the action diagram so that it gives a warning in such a case.

Answer

```
Entry: an exam mark
┌─IF (0≤mark≤39)
│   result = 'fail'
├─IF (40≤mark≤64)
│   result = 'pass'
├─If (65≤mark≤100)
│   result = 'distinction'
├─ELSE give warning
└─ENDIF
Exit: a grade or warning
```

Note that some authors would replace the second and third occurrences of IF with ELSEIF, to indicate more forcefully the fact that only one of the four possibilities may apply. But since the selection bracket indicates that only one choice is possible we can manage without ELSEIF.

Example 2
Verify that the action diagram below will carry out the correct grading of marks, provided the marks are valid; that is, in the range 0 to 100. Test it on a few typical marks.

```
Entry: a valid exam mark
┌─IF mark < 40
│   result = 'fail'
├─ELSE
│   ┌─IF mark < 65
│   │   result = 'pass'
│   ├─ELSE
│   │   result = 'distinction'
│   └─ENDIF-
└─ENDIF
Exit: a grade
```

Exercise 1.6.1
Expand the action diagram above so that a mark below 0 or above 100 produces a warning that the mark is out of range.

Repetition or iteration
The second of the action diagrams which we looked at earlier consisted of a sequence of five actions which a student had to carry out; these actions were: answer question 1, answer question 2 and so on. Now as far as we are concerned (the student may think otherwise!), this sequence is simply the five-fold *repetition* of the same action. This repetition (or iteration) can be described more economically if we introduce the repetition construct as shown below.

```
Entry: questions
┌─DO
│   answer the next question
└─UNTIL there are no more questions
Exit: answers
```

What we have is a loop: there is an instruction to carry out an action, (DO), repetitively, UNTIL a stopping condition occurs — so when there are no more questions to answer, we stop. Notice the convention that we have a double bar at the top of a repetition loop, and a single bar at the bottom.

There are two alternative ways of describing this iteration: the first alternative may look unnecessarily mathematical in this particular example, but it can be very useful in some circumstances.

>   Entry: questions
> $i=0$
>   ⎾DO WHILE $i<6$
> $i=i+1$
>   answer question number i
>   ⎿ENDLOOP
>
>   Exit: answers

In this case we have a sequence bracket which contains just *one* instruction, which is a DO — in other words a repetition. But notice that the DO instruction applies to a *sequence* of two instructions. The ENDLOOP is not essential, as the loop bracket shows us the extent of the loop, but it does emphasize the loop (in the way that we used an ENDIF construct for emphasis earlier).

A common alternative to this version of the repetition loop, which is also slightly shorter, is shown below. The instructions have exactly the same effect as the previous set of instructions.

>   Entry: questions
>   ⎾For $i=1$ STEP 1 TO 5
>   answer question number i
>   ⎿ENDLOOP
>   Exit: answers

In this case we have a counter, i, which has an initial value of 1, and which is increased by 1 each time the END LOOP is met. Before each repetition the value of i is checked, and for as long as it is between 1 and 5 inclusive a further repetition is carried out. After the fifth pass through the loop, the value of i will be 6, and no further repetition will take place.

Note that after the finish of a repetition loop — in this case after 5 repetitions — we continue to the next line below the loop, and carry out subsequent instructions.

Example 1
Suppose that in a card index (or more precisely a card *file*), each card is a *record* which contains details of the individual person whose name is written at the top of the card. The clerk who put the card file together did not bother to arrange it alphabetically, though he did put a blank green card on top of the index, and a blank red card on the bottom. Each of the cards relating to an individual person is white. Suppose we wish to read the information about a given person, to whom we give the name 'target'.

(a) Verify that the action diagram on page 22 will eventually provide us with the information about our target.

(b) What would be the consequence of there being no record corresponding to our 'target'?

```
Entry: card file, target
 ─examine next card
   ┌─IF card is not green
   │ print 'file is out of order'
   ├─ELSE
   │   flag=0
   │   ┌DO
   │   │ examine next card
   │   │   ┌─IF name on card = 'target'
   │   │   │  flag = 1
   │   │   │  copy details from card
   │   │   └─ENDIF
   │   └─UNTIL flag = 1
   └─ENDIF
 Exit
```

Notice that we've made a requirement that the first card in the file should be a green card. This is just one requirement for the file to be a valid file. Also we have introduced a 'flag' to tell us whether or not we've found the target record. The value of flag is initially 0, and is changed to 1 if and when the target record is located. In other words the value of flag determines whether or not to continue the iteration: if the value of flag is 0, the iteration continues. The simpler UNTIL condition

$\quad$ −UNTIL name on card = 'target'

might look initially more attractive: but we'll see below that the indirect condition — that is, the value of flag — is preferable.

Answer
(a) The instructions are to keep on looking for a card corresponding to target until the target value is found. As we said above, when the value of flag changes to 1, we know that the target has been located.
(b) We would keep on looking for a card corresponding to 'target' even though there are no cards to look at, since flag remains at 0. However, with our red card, we will be able to decide that we've reached the end of the file — and the state of the flag (1 or 0) will tell us whether or not we've successfully matched a card with the target. A possible upgraded version is shown below.

```
Entry: card file, target
 ─examine next card
   ┌─IF card is not green
   │  write 'error, card file out of order
   ├─ELSE
   │   flag = 0
   │   ┌DO
   │   │ examine next card
   │   │   ┌─IF card is red
   │   │   │  write 'error, target card does not exist'
   │   │   │  flag = 1
   │   │   ├─ELSE
   │   │   │   ┌─IF name on card = 'target'
   │   │   │   │  copy details from card
```

Algorithms and action diagrams

```
        │ │ │ │  flag = 1
        │ │ │ └ENDIF
        │ │ └ENDIF
        │ │ └UNTIL flag = 1
        │ └ENDIF
        └Exit:
```

Notice that as soon as it is clear that either the target card has been located *or* it does not exist, the value of flag is set to 1, and the iteration process terminates. Note that there is no ELSE associated with the third IF; in effect the latest card (which clearly is not red!) does not match 'target', so we proceed to examine the next card.

Example 2
Draw two alternative versions of the following action diagram

(a) using the DO...WHILE construction and
(b) using the DO...UNTIL construction.

```
    ┌X=0
    │ ┌FOR J = 1 STEP 1 UNTIL 5
    │ │ X=X+J
    │ └ENDFOR
    └PRINT X
```

Answer
One possibility for the DO...WHILE version is

```
    ┌X=0
    │ J=0
    │ ┌DO WHILE J<5
    │ │ J=J+1
    │ │ X=X+J
    │ └ENDLOOP
    └PRINT X
```

Notice that the final entry to the loop occurs when $J=4$. (How would this diagram have to be further modified if the two instructions within the loop had been in reverse order?)
A DO...UNTIL version is

```
    ┌X=0
    │ J=0
    │ ┌DO
    │ │ J=J+1
    │ │ X=X+J
    │ └UNTIL J=5
    └PRINT X
```

Compare this with the previous version.

Problems
1. (a) Produce an action diagram consisting of an iteration bracket of the FOR...UNTIL type to print all the numbers from 1 to 100 inclusive.
   (b) Modify the action diagram so that only the even numbers are printed.
2. The following rules are used to calculate weekly after-tax income for an employee. Income

below £50 is untaxed. Any income in excess of £50 has tax deducted at the rate of 20%. Draw an action diagram with pre-tax income as input and after-tax income as output, for a single employee.

3. Suppose that the people and animals that live on an island are specified by the characteristics listed below

Animal	No of legs	Colour	Horns?
pig	4	pink	no
man	2	N/A	no
cow	4	brown	yes
horse	4	brown	no
devil	2	green	yes
snake	0	green	no

On the assumption that there are no other species on the island, draw an appropriate action diagram to identify the name of the animal from its characteristics.

e.g.
```
┌ IF No of legs = 0
│   write 'snake'
└ ENDIF
```

What would the output of your action diagram be, from an input of the characteristics of a sheep (with four legs, grey colouring and no horns)?

2
Sets and set algebra

2.1 INTRODUCTION

In spite of the reassurance we have tried to give in Chapter 1, the title of this chapter probably confirms your worst fears about the sort of mathematics you are going to be studying. You may have experienced 'set theory' at school in the form of 'the set of all cows', 'the set of all animals that eat grass', and so on, in which case the subject may have seemed trivial, or useless, or both.

So before starting on the actual material of the chapter, we would like to explain why set theory is so fundamental to our subject as to merit first place in this book, and in particular what it has to do with computing.

The concept of a set is one of the most basic in the whole of mathematics. At present we will not give a formal definition of what we mean by the term, but will rely on intuitive understanding of a set as a collection of objects or items. This sounds extremely vague and general, but in fact it is for precisely this reason that the concept is so far-reaching and useful. By allowing the 'objects' in our sets to be of certain kinds, and later by introducing the idea of performing operations on the members of a set, we can unify apparently disparate branches of mathematics under the umbrella of set theory.

This unification has two benefits: first it helps us to establish links between areas of the subject which one might otherwise not suspect were connected, and second it means that we can, as it were, kill many birds with a single stone — results in different areas of the subject, which would otherwise require separate proofs, can be established by a single set-theoretic proof. So set theory actually saves work!

There is another way of looking at the idea of a set — and indeed, at the entire content of this book. We shall be defining and investigating the behaviour of increasingly complex 'mathematical objects' which can be used to reflect or 'model' complex computing problems. A set is in a way the starting point for this process — a collection of objects *without* any particular structure, on which later structures can be superimposed. In fact, later on in the chapter we shall be looking at two types of structured collection of set elements — n-tuples and lists.

Nevertheless, this concept of a set as a 'collection of objects' is probably not very helpful at first, so let us look at a few more concrete examples.

(i) When you were learning to count, you probably 'counted on your fingers' to begin with. You were using the set of your fingers as a kind of standard set, and you learned very early on that it had ten members for two hands, five members for one. In fact when you were counting beans or whatever at school, and numbered them off on your fingers 'one, two, three . . .' you were carrying out a set operation — setting up a direct correspondence between the members of the set of beans and the members of your set of fingers. If this direct correspondence could be established between your set of beans and the fingers of one hand, then your prior knowledge of that 'standard set' enabled you to say that there must be five beans.

Set theory is thus at the heart of the counting process; we will be returning to this idea in Chapter 4, on data representation and manipulation, and in Chapter 9, on methods of counting and enumeration.

(ii) When we are concerned with describing mathematical structures, or the real-world structures which they represent in a way which can be handled by a computer, set-theoretic terminology often provides a convenient way of doing so. For example, if you are trying to solve a problem concerning the most efficient way of transporting goods around over a road network, you will need a way of describing that network. Now for a human being to give such a description to another human being is not too difficult — you can draw a map, or say things like 'Well, there are four towns, Avonmouth, Bristol, Chippenham and Devizes, but you can only get from Avonmouth to Chippenham by going through Bristol . . .' and wave your hands. However, this kind of human skill is not matched by the current generation of computers — we need to be much more precise and formal about the way we express the problem.

For example, you could describe the road network by deciding to represent each town by the first two letters of its name (why *two* — well, if we only used the first letter, what would happen if Bath joined our list of towns?). Then we could write the set of towns under consideration as {Av, Br, Ch, De} — don't worry too much about the set notation at this stage, it is the idea which is important. The set of roads linking the towns can then be described by a set of pairs of towns {{Av, Br}, {Ch, De}, . . .}, so that there is a relationship between the two sets. Moreover, having once defined the sets we can then start asking questions like 'what properties must the two sets have if it is to be possible to get from any town to any other town without passing through a third?' So a knowledge of relations between sets, and of special properties of sets, is going to be necessary.

We will be returning to this kind of problem in Chapter 5, on graphs.

(iii) Finally, an example which is more directly related to computing. Many important applications of computing at the moment involve databases, and set theory can provide a very compact way of describing the various categories of information which may be stored on a database. For instance, a single file in an organization's database may contain the set of all names of employees; another may contain the set of all department codes within the organization, and so on.

Once again, the question of the relations between these sets will arise — there is clearly a relationship of the form 'belongs to' between the members of the employee set and those of the department code set. But other possibilities are opened up here. Structures or orderings can be imposed upon the members of the sets — 'reports to', for example. We will take a brief look at this kind of relation later in the chapter, though a fuller discussion will have to wait till Chapter 8.

We may also wish to query the database, so that we need to be able to link together statements about the members of the various sets in a logical way: 'If X is in the accounting department and X earns less than £8000 per annum then X must be a Grade B Accounts Clerk'. So a consideration of sets of this kind leads naturally to the application of elementary logic, which is the subject of the next chapter.

2.2 SETS

First, some definitions.

A *set* is a collection of items (called the *elements* of the set) defined in such a way that for any given item, it is possible to say whether the item is or is not contained in the set.

This may seem to be long-winded, but the fact that elements can be unambiguously assigned to sets is actually very important. Though there is now growing interest in a kind of mathematics in which the boundaries of sets are 'fuzzy', so that some items are neither definitely in nor definitely outside the set — a situation which mirrors some real-life circumstances very well — it is still true that, for the great majority of set applications, this clear-cut boundary to the set is realistic. For example, if we are discussing the set of all products manufactured by a company, then it will be known, for any particular product, whether or not the company makes it — there is no intermediate position.

As we have already pointed out, the question of notation — efficient ways of writing mathematical statements which can be manipulated easily and, often, translated into computer-readable form — is an important one. So we need to begin with a neat way of stating that an element does, or does not, belong to a set. Generally we use ordinary capital letters to denote sets — A, S, and so on — and the symbol $\in$ to mean 'belongs to' or 'is an element of'. Thus $x \in A$ is to be read as 'x belongs to the set A', while $y \notin B$ means 'y is not an element of the set B'.

The immediate consequences of our definition of a set is that for any set we need a way of describing the elements of the set. This can be done in a number of ways.

(i) If the set has only a small number of elements, we can list them individually, enclosing them in brackets, usually of the form { }. Thus we could write {England, Wales, Scotland, Northern Ireland} to denote the set of countries which together make up the United Kingdom. This has the advantage of being simple and unambiguous, but it will obviously get very cumbersome for large sets.

Sometimes, if the list of elements has a pattern which will be clear from the first few, we omit some of the elements and use the symbol ... to show that they have been omitted. For instance, if we wish to define a set consisting of all multiples of three between 1 and 29, we could write {3, 6, 9, 12, ..., 27}.

(ii) Alternatively, we can define a set by stating some property of its elements; for instance, the set of multiples of 3 described above could be written as {all multiples of 3 between 1 and 29}. In this way we can also define sets which have an infinite number of elements, and therefore could not be listed explicitly, such as {all even integers} or {all powers of 2}.

Because these descriptive sentences can get rather cumbersome, a version known as *predicate form* is often adopted. It will be easier to see how this operates by looking at a specific example.

$$S = \{x : x = 3n, n \text{ an integer, and } 1 < x < 29\}.$$

This is actually a definition of the same 'multiples of three between 1 and 29' set we defined above. The statement can be read as 'S is the set of elements x such that x has the form 3 times an integer, and x lies between 1 and 29'. In particular, the : is read as 'such that', a piece of notation which you will come across again later in the book.

It is worth noting here that, whatever mode we choose for defining the elements of a set, the order in which they occur does not matter at all — the set {2, 4, 6} and the set {6, 2, 4} are identical. Later in the chapter we will encounter situations where ordered sets are important, but in general, unless otherwise stated, the order of set elements is of no significance. We may choose to write them in a certain order, such as {2, 4, 6, ...}, for convenience, particularly if we want to make clear the pattern of elements in an infinite set, but any other order would give us precisely the same set.

Before going any further, test your comprehension of the concepts introduced thus far with the exercise below.

Exercise 2.2.1
Which of the following do you think constitute valid definitions of sets? Where possible, list the elements of the set in question. (They are not intended to be watertight, so you may argue about some of them.)

(a) The set of all possible routes from London to Bristol by public roads.
(b) The set S such that $x \in S$ if x is an integer and $0 \leq x \leq 20$.
(c) The set of valid statements in PASCAL (or another computer language with which you are familiar).
(d) The set of all domestic animals.

2.3 FURTHER SET CONCEPTS

We looked above at ways of defining the elements of a set, but there is one important set whose characteristic is that in fact it has *no* elements. This set is called the *empty set*, and is usually written as $\emptyset$, or sometimes as { } — a set that is literally empty. One point worth noting is that {0} is *not* the empty set, since it has one element, namely 0. The number of elements in a set, S, often called its *cardinality*, is written $|S|$. So $|\emptyset| = 0$.

Often we want to deal with sets, the elements of which are also elements of a larger set. For instance, the elements of the set $A = \{$all students on the first year of the HND Computer Studies course} will be contained in the larger set $B=\{$all students on the HND Computer Studies course}. We then say that A is a subset of B, and write $A \subseteq B$. Formally we can say that A is a subset of B if given any element x such that $x \in A$, it follows that $x \in B$ also.

The empty set is, of course, a subset of all other sets, even though we cannot, as it were, 'see' it in the set. Perhaps more surprisingly, we regard any set as a subset of itself, $A \subseteq A$, since the definition above is certainly satisfied. A set $A \subseteq B$ which satisfies the definition of a subset, but for which it is possible to find an element $y \in B$ such that $y \notin A$, is called a *proper subset* of B, also written as $A \subset B$.

Sometimes it is convenient to regard all the sets we are considering as selected from one overall larger set. For example, in the case of the sets of all first-year HND Computer Studies students, and all HND Computer Studies students introduced above, both these sets could be regarded as subsets of the larger set $X=\{$all students on full-time courses at this institution}. The set X is then referred to as the *universal set*. However, you should be clear that there is no such thing as *the* universal set — how it is defined will depend on the particular problem we are considering.

Finally in our definitions of basic properties of sets we need to consider what equality of two sets means. Intuitively it is clear that if we say $A=B$ we mean that A and B have identical elements, but often this is not easy to prove. A better definition is that $A=B$ if $A \subseteq B$ and $B \subseteq A$ — all elements of A are also elements of B, and vice versa.

At this point we should perhaps have a brief look at the idea of Venn diagrams. If you 'did sets' at school you probably spent quite a lot of time over these. Basically a Venn diagram is a way of showing a set, or relations between sets, in pictorial form; for example, Fig. 2.1 shows the relation $A \subset B$.

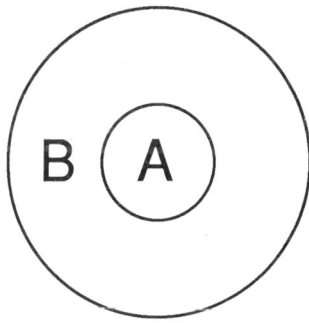

Fig. 2.1.

These diagrams can be useful in clarifying your ideas about the relation between sets, or in getting some feeling about how to tackle a particular problem. Their importance, however, should not be exaggerated; they can never constitute a proof of a result, and in some circumstances, for instance when a set is infinite, they are really no use at all. So how much use you choose to make of them will be left very much up to you.

Exercise 2.3.1
A set can have elements which are themselves sets — for example, the set of classes in a school has members which are sets of pupils. Consider the following set: $S=\{\{a\}, \{c\}, \{a\,b\,c\}, \{\emptyset\}\}$. Which of the following statements are true?

$\{a\} \in S$
$\{a\} \subseteq S$
$b \in S$
$\{b\} \in S$
$\{\emptyset\} \in S$
$\emptyset \in S$

Exercise 2.3.2
Define sets as follows:

$N = \{$all positive integers and zero$\}$ (an integer is a 'whole number', such as 2, or 1234).
$S = \{x : x=n^2, n \in N\}$
$L = \{0, 2, 4, 6, 8, \ldots\}$
$E = \{$all positive multiples of 8$\}$
$M = \{y : y=2k, k \in N\}$

(a) Which, if any, of these sets are equal?
(b) Identify all pairs of sets $[A, B]$ in the list for which the statement 'A is a subset of B' is true. (Notice that the order in which we write $[A, B]$ is important here — $[A, B]$ is not the same as $[B, A]$. This contrasts what we have said about the listing of set elements. In fact $[A, B]$ is an example of an *ordered pair*, of which you will be hearing more later).
(c) Describe in words the set Q defined by the following:

$Q = \{y : y=a/b, a, b \text{ integers}, b \neq 0\}$

2.4 SOME SPECIAL SETS

There are certain sets which crop up so frequently that they have been given standard names for reference. The ones we will be using are as follows.

(a) The set of *integers* or whole numbers. This is written as Z, so $Z = \{\ldots, -2, -1, 0, 1, 2, 3, \ldots\}$.
(b) The set of *rational numbers*, Q. A rational number is one which can be expressed as the ratio of two integers. So $Q = \{x : x=p/q, p, q \in Z, q \neq 0\}$.
(c) The set of natural numbers $N = \{0, 1, 2, 3, \ldots\}$
(d) The set of real numbers R — this includes positive and negative integers, fractions, and also numbers such as π which cannot be written exactly either as fractions or as recurring decimals.

2.5 OPERATIONS WITH SETS

So far we have had a lot of definitions *about* sets, but we have not *done* very much with them. There are four ways of combining the elements of sets into new sets, and we will consider each in turn, then look at the kind of 'set arithmetic' which can be built up using all four operations.

(i) Union

The *union* of two sets A and B is written as $A \cup B$, and consists of the set containing all the elements of both A and B. Thus if $A = \{2, 4, 6\}$ and $B = \{3, 6, 9\}$ then $A \cup B = \{2, 3, 4, 6, 9\}$. A Venn diagram showing this operation is given in Fig. 2.2.

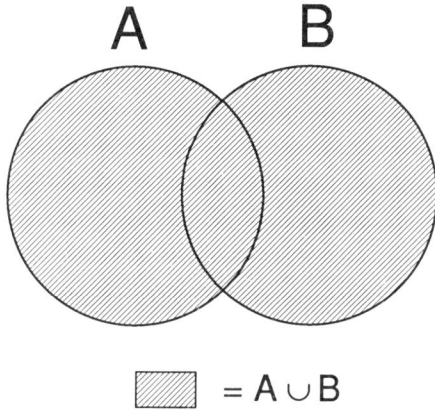

Fig. 2.2.

(ii) Intersection

The *intersection* of two sets A and B, written $A \cap B$, is the set containing all elements which belong to *both* A and B. For the sets A and B defined in (i) above, $A \cap B = \{6\}$. If A and B have no elements in common then of course $A \cap B = \emptyset$, and A and B are said to be *disjoint*. A Venn diagram for the intersection operation is shown in Fig. 2.3.

It is clear from the Venn diagrams that $A \cup B = B \cup A$, and $A \cap B = B \cap A$ — in other words, that both union and intersection are *commutative* operations. We are, of course, used to this with some of the operations of 'ordinary arithmetic', such as addition and multiplication, but later on in the book you will come across operations which are *not* commutative, and so it is a good idea to get out of the habit of taking this for granted. In fact, it is not true even of such simple operations as division or subtraction.

(iii) Difference

The *difference* of two sets is denoted $A - B$, and consists of all the elements of A which are not in B. Symbolically, if $C = A - B$ then $x \in C$ if $x \in A$ but $x \notin B$. Of course, if A and

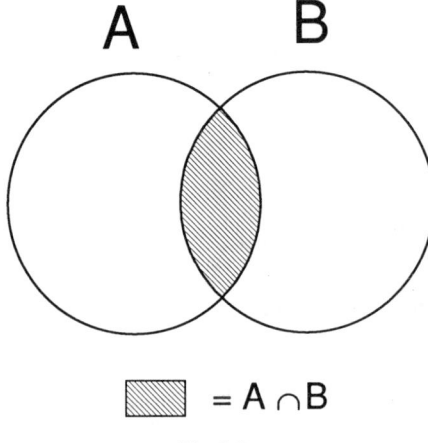

Fig. 2.3.

B are disjoint then $A-B=A$. For the sets A and B defined above, $A-B=\{2, 4\}$, while $B-A=\{3, 9\}$. Fig. 2.4 illustrates this operation.

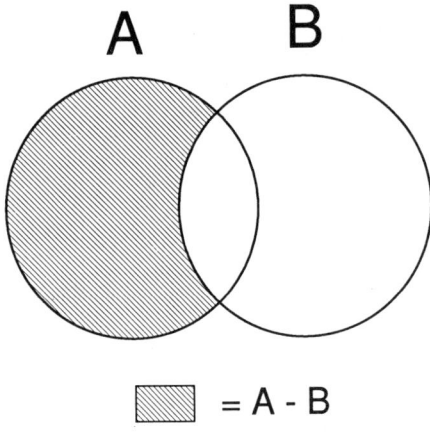

Fig. 2.4.

(iv) Complement
The *complement* of a set, sometimes written $\bar{A}$ or A', consists of all the elements of the relevant universal set which are *not* contained in A; another way of expressing this is that $A'=U-A$. Thus $x \in A'$ if $x \in U$ and $x \notin A$. If, for the set $A=\{2, 4, 6\}$ which we have been using, the universal set in question is the integers from 1 to 10, then $A'=\{1, 3, 5, 7, 8, 9, 10\}$. Fig. 2.5 illustrates the idea.

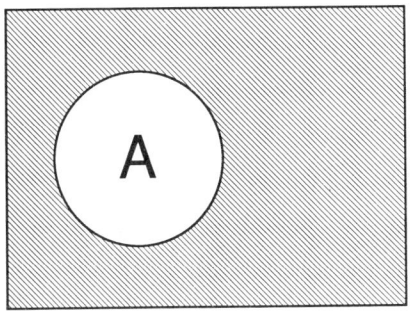

Fig. 2.5.

Example
Let U = the set of all students in a college,
      F = the set of female students
      M = the set of male students
      C = the set of students who study computing
      S = the set of students who study statistics

(i) Interpret verbally F', $U-C$, $F \cap (U-C)$
(ii) Express symbolically the set of all male students studying statistics; the set of students studying both statistics and computing; the set of female students who study either statistics or computing, but not both.

Answer
(i) F' = the set of all non-female students — which of course is M. $U-C$ = the set of all students who do not study computing. $F \cap (U-C)$ = the set of female students who do not study computing.
(ii) $M \cap S$; $C \cap S$; $F \cap [(S \cup C) - (S \cap C)]$. The last one is a bit complicated — Fig. 2.6 may help. Can you spot another way of writing the expression in square brackets, which represents 'students studying statistics or computing but not both'?

This last example really involves what we might call the 'algebra' of set operations — how the different operations can be combined, and what rules must be obeyed in combining them. We return to this topic in the next section.

Exercise 2.5.1
A polytechnic has three departments — Science, Computing and Mathematics — at its Greenfield site, two departments — Business Studies and Accounting — at its Fairfield site, and three — Engineering, Design and Languages — at its Hilltop site. Each department has its own set of computer terminals; the terminals belonging to Science will be denoted by the set S, the set of terminals at the Greenfield site by G, and so on. The universal set here is the set of all terminals owned by the polytechnic. Describe in words each of the following sets:

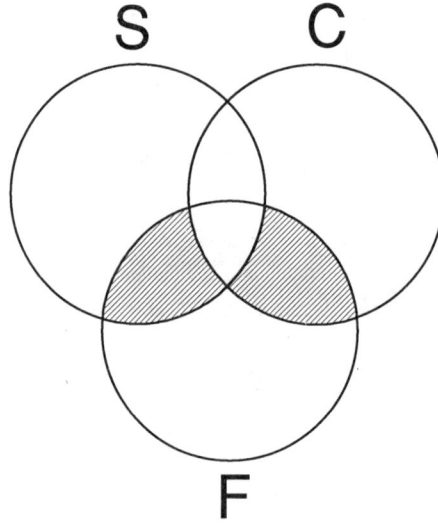

Fig. 2.6.

(a) $(S \cup M) \cup C$
(b) $(B \cup A)'$
(c) $F \cup H$
(d) $H - (D \cup L)$

Exercise 2.5.2
In a company the set of members of the Data Processing division is denoted by D, the set of employees earning more than £12000 per year by R, and the set of employees aged 35 or over by S. Write expressions in set-theoretic notation to represent the following:

(a) The set of employees aged less than 35.
(b) The set of employees under 35 years old earning more than £12000 per year.
(c) The set of employees who are *either* members of the DP department, *or* earning more than £12000 a year, but not both.
(d) The set of employees who are members of the DP department aged under 35 years.

2.6 THE ALGEBRA OF SET OPERATIONS

Let's begin by reminding ourselves of some of the facts about ordinary arithmetic which we generally take for granted. We have already mentioned the *commutative* property of addition and multiplication — the fact that $2 \times 3 = 3 \times 2$, and $4 + 7 = 7 + 4$. The order in which the numbers are added or multiplied makes no difference to the answer. We have also seen that set unions and intersections are similarly commutative.

There are two other important laws in arithmetic: firstly the *distributive* law, which tells us that $a \times (b+c) = (a \times b) + (a \times c)$, in other words that multiplication is distributive over addition. Of course, it is *not* true that addition is distributive over multiplication — this would imply that $a + (b \times c) = (a+b) \times (a+c)$, which is untrue, as

Sec. 2.6]	**The algebra of set operations**	35

substitution of a few simple numbers will confirm. Secondly we have the *associative law* which says that $(a\times b)\times c=a\times(b\times c)$, and $(a+b)+c=a+(b+c)$.

Finally in the arithmetic of ordinary numbers there are a couple of rather special numbers: zero, which has the property that $a+0=a$, and one, which is such that $a\times 1=a$, both relations being true for any number a. We characterize these two numbers by saying that 0 is the *identity* (or sometimes *identity element*) for addition and 1 for multiplication. Moreover, having defined these identities, we can go further and define *inverses*: the additive inverse of a, usually called the negative, is $-a$, the number which when added to a gives us zero, while the multiplicative inverse of a, written $1/a$, is the number which when multiplied by a gets us back to 1. (Note that zero itself does not have a multiplicative inverse.)

So to sum up, the operations of ordinary arithmetic have the properties that

(i) addition and multiplication are commutative
(ii) multiplication is distributive over addition
(iii) both operations are associative
(iv) there are identity elements for the operations of addition, written as 0, and multiplication, written as 1.
(v) any number a has an additive inverse, written $-a$, such that $a+(-a)=0$; and any non-zero number a has a multiplicative inverse, written $1/a$, such that $a\times(1/a)=1$.

What we will do now is investigate which, if any, of these properties also apply to the algebra of sets, and whether there are any other properties, not found in ordinary arithmetic, which apply in set algebra, We can establish an analogy — which should not be pushed too far! — between union and addition, and between intersection and multiplication. In a sense when we take the union of two sets we are 'adding' their elements together. We already know that both operations are commutative, so the commutative law also applies to sets under both union and intersection. But what about the other properties?

Example
Verify that $\cap$ is distributive over $\cup$, i.e. that

$$A\cap(B\cup C)=(A\cap B)\cup(A\cap C).$$

Answer
This can be seen to be true by examining the corresponding Venn diagram (Fig. 2.7). (Note however that this does not constitute a formal proof of the result.)

If you try to describe to yourself in words the Venn diagram in Fig. 2.7, you will probably find that the word 'or' keeps cropping up in connection with union, the word 'and' in connection with intersection; for instance, the elements of $A\cap B$ are those which are in A *and* B. In Chapter 3 we will investigate this connection more fully.

Exercise 2.6.1
Verify by a similar argument that $\cup$ is also distributive over $\cap$, i.e. that

$$A\cup(B\cap C)=(A\cup B)\cap(A\cup C)$$

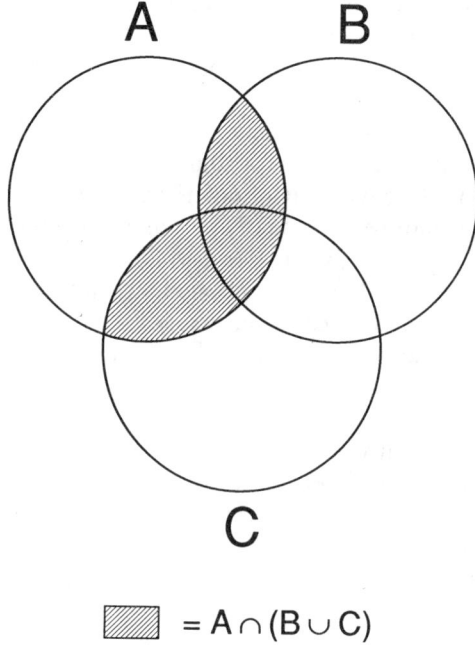

$$\text{▨} = A \cap (B \cup C)$$

Fig. 2.7.

Exercise 2.6.2
Produce a Venn diagram, and also a verbal/symbolic argument, to verify that the associative law holds for both union and intersection, that is, $A \cup (B \cup C) = (A \cup B) \cup C$, and $A \cap (B \cap C) = (A \cap B) \cap C$.

To find the analogue of the number zero in ordinary arithmetic, we must consider what properties we wish this 'zero set' to have. The number zero is such that $a+0=a$ for any number a, so we are looking for a set — let's call it X for the moment — which is such that $A \cup X = A$. As you may have guessed already, the only set which has this property is the empty set; so $\emptyset$ with $\cup$ in set algebra takes the role of 0 with $+$ in arithmetic. (It is of course also true that $A \cup A = A$, but we cannot identify this A with the X we are looking for; X must be the *same* for *all* A.)

Exercise 2.6.3
Can you suggest what set I has the property, analogous to that of 1 in arithmetic, that $A \cap I = A$ for any set A?

We have now verified that identities exist for both union and intersection.

Example

Is there anything in set theory corresponding to the negative in ordinary arithmetic — that is, is there a set X with the property that $A \cup X = \emptyset$, the 'union' identity?

Answer

A little thought should convince you that there is in fact *no* set X with this property. Likewise, if we try to find a set Y such that $A \cap Y = U$, we will hunt for a long time.

However, we *can* make two rather similar statements, connected as you might expect with the one set operation we have not made use of yet, namely the complement. Recall that $A' = U - A$; then it follows that $A \cup A' = U$ and $A \cap A' = \emptyset$.

We can thus sum up the properties of set algebra which we have established, as follows:

1. $A \cup B = B \cup A$ and $A \cap B = B \cap A$.
2. $A \cup (B \cup C) = (A \cup B) \cup C$ and $A \cap (B \cap C) = (A \cap B) \cap C$.
3. $A \cap (B \cup C) = (A \cap B) \cup (A \cap C)$ and $A \cup (B \cap C) = (A \cup B) \cap (A \cup C)$.
4. $A \cup \emptyset = A$ and $A \cap U = A$.
5. $A \cup A' = U$ and $A \cap A' = \emptyset$.

You may be feeling that we have made rather a meal of establishing these 'rules' of set algebra. We have done so for two reasons: first, to demonstrate that the kind of 'commonsense' rules we are accustomed to using in arithmetic cannot always be assumed to carry over to the kinds of mathematics which are useful in computing; secondly, we will find in the next chapter that a set of rules very similar to those we have just derived applies also in the context of logic — and in fact to a whole group of 'algebras' of great generality and importance.

2.7 SELECTIONS FROM SETS

In many applications of set theory, we are concerned with the process of selecting items, or subsets, from a given set according to certain rules. For example, in probability theory we often need to count the number of ways in which an event can occur: how many ways are there for a message to be transmitted through a computer network? how many different samples of 10 computer users can we select from a firm in which there are 50 users? In this section we will look at a number of ways in which such selections can be defined.

First, we consider the set of all possible subsets of a given set. We have already tackled problems in which we identified subsets of a given set. Now suppose we turn the problem round, and ask, 'How many subsets, of all sizes, does this set have?' Remember here that we established a convention that the empty set is a subset of all other sets, and that any set is a subset of itself.

This 'set of all subsets' of a set S is called the *power set* of S, and for small sets it is easy enough to write out the power set in full. For instance, if $S = \{a, b, c\}$ then its power set will have elements $\{\ \}, \{a\}, \{b\}, \{c\}, \{a, b\}, \{a, c\}, \{b, c\}, \{a, b, c\}$ — eight elements in total. The power set, like the set you encountered in the exercise at the end of Section 2.3, is a set whose elements are themselves sets.

Exercise 2.7.1
Find by enumeration (that is, by writing the subsets out) the power set of $S = \{a, b, c, d\}$. Can you generalize by looking at this and the three-element set discussed above to get a formula for the number of elements in the power set of an n-element set? (We will be proving this result properly later in the book).

Example
A nice illustration of the idea of a power set is provided by the set of possible characters in Braille. A 'cell' of Braille consists of a pattern of up to six raised dots, arranged as shown in Fig. 2.8.

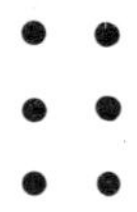

```
• •
• •
• •
```

Fig. 2.8.

We can regard this cell as a set C of six possible dot positions. The set of all possible characters is then the power set of C — the set of all possible subsets selected from C. By the result you obtained in the exercise above, you should be able to see that there will be 2^6 such subsets. However, this includes the empty subset, which is a blank character, so in fact there are $2^6 - 1$ or 63 possible Braille characters. In this fact lies one of the strengths of braille — as only 26 of these characters are used up in representing the letters of the alphabet, plus a few for punctuation marks, the remainder can be used to represent common words such as 'and' and 'the', endings like '-tion' and 'ing', and so on. These contractions, as they are called, help to reduce the volume of Braille required in transcribing a piece of printed matter.

2.8 ORDERED PAIRS, RELATIONS AND CARTESIAN PRODUCTS

There are other ways of selecting set elements besides choosing all subsets. We might specify subsets with a certain number of elements — all three-element subsets, for instance. This idea is taken one stage further in the concept of an *ordered pair*, which is a two-element subset *in which the order of selection matters*. To avoid confusion, we usually write ordered pairs inside square brackets, thus: $[a, b]$.

Much useful information can be conveyed by the use of ordered pairs. In the road network problem introduced in Section 2.1, we could denote travel from Bristol to Chippenham by the ordered pair [Br, Ch] — not, of course, the same thing as [Ch, Br], which would represent travel in the reverse direction. In the original example, where we simply wanted to specify the existence of a road link between two places, and not the direction of travel, we wrote {Ch, Br} to indicate that this was simply a set of two linked towns, with no particular order implied.

We can also use ordered pairs as a neat way of expressing relations between set elements. Suppose we are considering the set of five employees of a business whose relationships are illustrated by the extract from an organization chart shown in Fig.

2.9. (This chart is actually an example of a *tree*, a structure mentioned briefly in Chapter 1, and to which we will be returning in Chapter 7.)

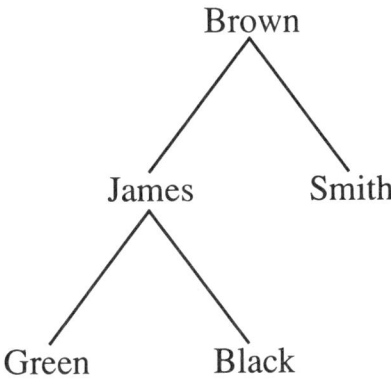

Fig. 2.9.

If we define a relation 'is subordinate to', and another relation 'is superior to', we can then write [x, y] to mean 'x is superior to y'. So (Brown, James) means Brown is superior to James. Try writing down the other members of this relation. You can see that the entire relation is itself a set — a set of ordered pairs. Now denote 'a is subordinate to b' by [a, b] and write down the set of ordered pairs which constitutes this relation. What do you notice about this set as compared to the previous one? We say that these two relations are *inverses* of each other — if a is superior to b, then b is subordinate to a. There will be more about inverse relations later in the book.

Actually there is no reason why the elements of an ordered pair need be chosen from the same set. If we have a set P consisting of all parents of children at a school, and another set C consisting of all children at the school, then clearly a relation 'is a parent of' could be defined, and shown as a set of ordered pairs each of which contains a first element selected from P and a second from C. Thus (James Brown, Mary Brown), where James Brown $\in P$ and Mary Brown $\in C$, is a member of the set representing the relation, and is interpreted as 'James Brown is a parent of Mary Brown'.

In the above example, our relation would not consist of all pairs of elements taken one from P and one from C — only those pairs for which the relation actually holds. But there are many cases in which we *do* want to consider the set of all possible pairs made up of an element from set X and one from set Y. We define this set to be the *product* of X and Y thus: $X \times Y$. So if $X = \{a, b, c\}$ and $Y = \{p, q\}$ then $X \times Y = \{[a,p], [a,q], [b,p], [b,q], [c,p], [c,q]\}$.

One familiar example of this situation is the way we are accustomed to specify points in the two-dimensional plane of a graph by their co-ordinates [x, y], where x and y are real numbers. Looked at in this way, the set of all possible co-ordinates for points is the product of the set of all real numbers with itself. Since co-ordinates of this kind are called Cartesian co-ordinates, after the French mathematician Descartes, the product of two sets defined in this way has come to be known as the *Cartesian product*.

We can also consider sets of ordered pairs which define subsets of the plane, rather than the whole plane. For example, the set $\{[x, x^2]:x\in R\}$ would represent the graph of $y=x^2$; the set $\{[x, y]:0\leqslant x\leqslant 1 \text{ and } 0\leqslant y\leqslant 1, x, y\in R\}$ would represent the area shaded in Fig. 2.10, and so on. This is how graphics 'windows' are often described, though because of of the discrete nature of a graphics screen, we would have $x\in N$ rather than $x \in R$.

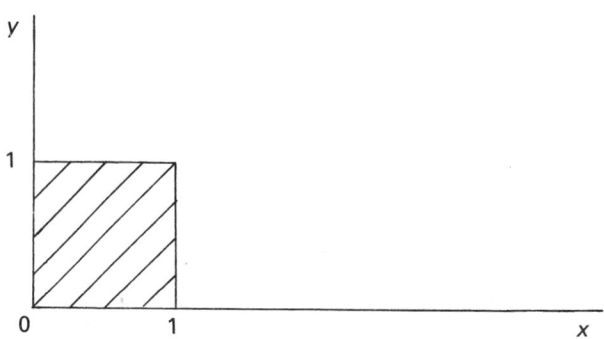

Fig. 2.10.

In the case of the set $\{[x, x^2]:x\in R\}$, there is a sort of rule or recipe which enables us to find the second element of each pair, given the first. In this case it is, of course 'square x', a rule which will get us from the values of x — that is, all the elements of the set R — to the values of x^2. However, x^2 cannot have absolutely any value in R — it must be positive. So our 'rule' takes us from the whole of R to a subset of R.

Example
Write an ordered-pair description of (a) the graph of $y=1/x$, (b) the region shown in Fig. 2.11(a), (c) the region below the line $x+y=2$ and lying in the positive quadrant — see Fig. 2.11(b).

Answer
(a) The set is $\{[x, 1/x]:x\in R, x\neq 0\}$. 0 has to be excluded since 1/0 is not defined (unless of course we extend our idea of R to include infinity).
(b) The defining property of the region is that x lies between 1 and 2, and y between 2 and 4. So we can say that points within the region are $\{[x, y]:x, y\in N; 1\leqslant x\leqslant 2, 2\leqslant y\leqslant 4\}$. If we used the $<$ sign rather than $\leqslant$ then points on the boundary of the region would not be included.
(c) $\{[x,y]:x\in R, 0\leqslant x\leqslant 2, x+y\leqslant 2, y\geqslant 0\}$.

2.9 N-TUPLES

Of course, there is no reason to limit ourselves to ordered pairs — we could define an ordered collection of any number of set elements. In general an ordered collection of n elements is called an n-tuple. Thus if we have an alphabetic arrangement of the 32 members of a school class, this could be regarded as a 32-tuple, and written $[m_1, m_2,$

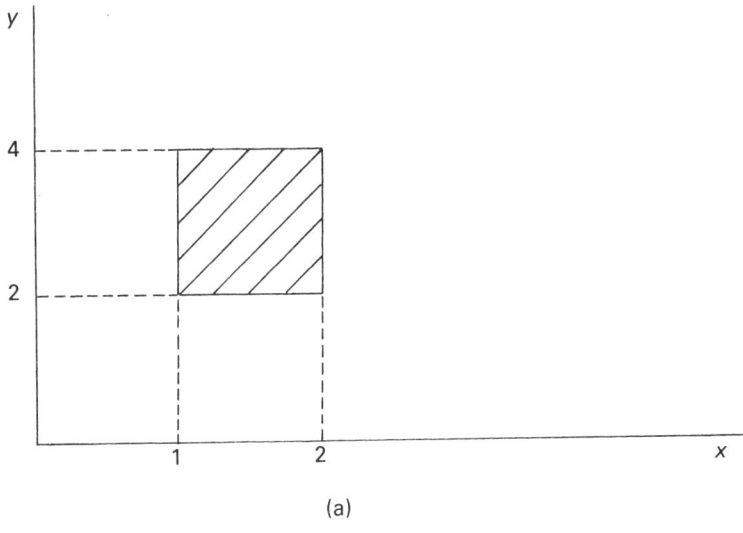

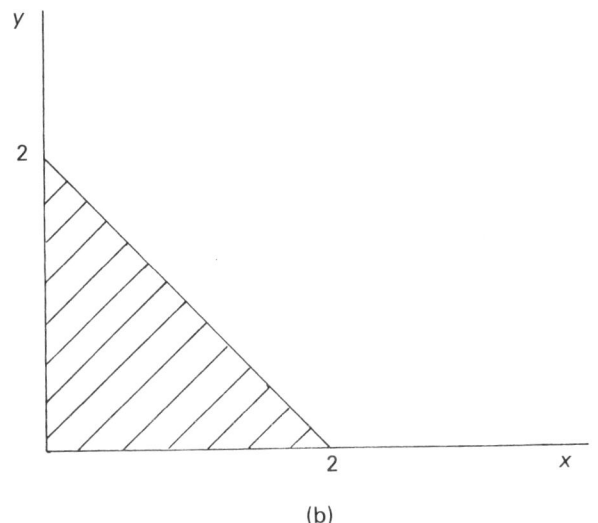

Fig. 2.11.

..., m_{32}], where m_1 denotes the first member, and so on. The numbers 1, 2, ... here are called *subscripts*, and we could write the whole 32-tuple in a more compact way as $[m_k, k=1, 2, \ldots, 32]$. Note that, as with ordered pairs, we use round brackets for n-tuples.

The idea of an n-tuple may already be familiar to you from your computing experience, under the alternative name of an *array*. The 32-tuple described above is a one-dimensional array, since one single subscript is sufficient to define uniquely the position of the element within the array. We usually call it a 1×32 array. Frequently instead of using subscript notation we write the array parameter within brackets; thus the elements of the 32-tuple could be denoted by $m[k]$, $k=1, 2, \ldots, 32$. In Chapter 4 we will be looking in more detail at two-dimensional arrays and ways of manipulating them.

The idea of an array or an n-tuple is fine as long as we know exactly how many elements we want to include in our ordered collection. But when it comes to computer storage of this kind of information, there can be difficulties, arising mainly from the need to know the size of the n-tuple in advance. Suppose, for example, that in the case of the 32-tuple defined above an extra student joins the class late. The 32-tuple now becomes a 33-tuple (or the 1×32 array a 1×33 array), and any references we have made to the size of the array in programs designed to operate on it will need to be amended accordingly.

We therefore need an alternative kind of structure to cope with possible variations in the sizes of our n-tuples — one which preserves the property that the order of elements matters, but in which n does not need to be pre-determined. Such a structure is provided by the idea of a *list*, and we will examine this concept in the next section. First, however, a couple of exercises on ordered selections to increase your familiarity with the ideas.

Exercise 2.9.1
A team of climbers tackling a Himalayan mountain is split up over six camps. The camps are connected by radio links, but owing to differences in equipment Camps I, II and III can both receive and send messages, while Camps IV and V can only send them; the team at Camp VI has lost its radio equipment altogether. Write out the set of ordered pairs representing the relation 'can receive a message from', and the set for 'can send a message to'.

Exercise 2.9.2
List the members of the set $M\times N$, where $M=$ the set of positive integers less than 6, and $N=\{0, 1\}$.

2.10 LISTS AND LIST OPERATIONS

As we saw at the end of the last section, it can be a nuisance having to specify the size n of an n-tuple in advance, especially if this size is likely to alter in time. Accordingly we define a *list* to be a collection of set elements which are arranged in a specific order, but with the number of elements involved being variable and unspecified.

To distinguish a list from an n-tuple, we write it inside pointed brackets. Thus $\langle$Newport, Cardiff, Bridgend, Neath, Swansea$\rangle$ is a list of stations on the main British Rail South Wales line. So we now have three notations to remember: $\{a, b, \ldots\}$ for sets, $[a, b, \ldots]$ for n-tuples, and $\langle a, b, \ldots \rangle$ for lists. It is important to get the distinction between these three clear in your mind; for instance, the sets {Bridgend, Neath, Swansea} and {Neath, Bridgend, Swansea} are identical, but the lists $\langle$Bridgend, Neath, Swansea$\rangle$ and $\langle$Neath, Bridgend, Swansea$\rangle$ are not, since the order is different in the two cases.

There is another advantage which a list possesses as compared with an n-tuple — it is possible to have an empty list, whereas an 0-tuple is difficult to envisage. The empty list, like the empty set, is denoted by the brackets with nothing between them

⟨ ⟩. This is a matter of more than theoretical interest in computing. Suppose, for example, that you are processing invoices for a large number of customers, and wish to know when all customers on your list have been invoiced; one way of finding out when this is the case is to store a list of all *remaining* customers — those who have *not* been invoiced — and test after each invoice has been issued to see whether this list is empty. When it is, all customers have been dealt with. This method does not require us to know in advance how many customers are on a file, and so is superior to possible alternatives such as counting the customers as they are invoiced.

This last example introduces the idea of list processing; a great deal of commercial computing consists of this kind of activity, and not only in the obvious sense of processing lists of orders, employees, and so on. A piece of text on a VDU (visual display unit) screen can be regarded as a list — of lines, words, or characters. In fact the text can be considered as a list of lists — a list of lines, which are themselves lists of words (which, for that matter, are lists of characters). So can you see that the idea is quite far-reaching.

There are two further definitions related to lists which are important in list processing: the *head* of the list — that is, the first item, or the item at the top, or front, of the list, and the *tail*, which is everything else in the list *except* the head. Thus in our list of stations ⟨Newport, Cardiff, Bridgend, Neath, Swansea⟩, Newport is the head of the list, and the tail is ⟨Cardiff, Bridgend, Neath, Swansea⟩.

You can see immediately that the head of the list is a single set element, whereas the tail is itself a list. We can define three processes related to these definitions: head (list), which selects the head element from a list, tail (list) which selects the tail, and append (list1, list2), which produces a new list whose elements are those of list1 in order followed by those of list2 in order. These ideas will become clearer if we look at an example.

Example
Suppose list1=⟨Adams, Brown, Carter⟩, and list2=⟨Drew, Evans, Finch, Green⟩. We will evaluate various expressions involving heads, tails, and appends.

append (list1, list2)=⟨Adams, Brown, Carter, Drew, Evans, Finch, Green⟩.

head (append (list2, list1))=Drew.

tail (append (list2, ⟨Heath⟩))=⟨Evans, Finch, Green, Heath⟩.

tail (tail (list2))=⟨Finch, Green⟩.

If you cannot follow any of these merely by inspection, you should work through actually writing out the lists at each stage.

Armed with these three quite simple operations on a list, many forms of complex list-handling can be carried out. We will not pursue the topic of list-handling any further here, but will be returning to the ideas in later chapters. For the present what is important is that you remember the terminology and its meaning.

A word of warning may be in order here. There is some ambiguity in the names used to describe lists and associated structures. Some authors reserve the name 'list' for an ordered collection of set elements of which *only* the head can be accessed;

others refer to *any* ordered collection of set elements as a list, and call a list of which only the head is accessible a 'stack' — a term you may have come across in your computing already. There are other similar structures in use, too: a *queue*, for example, is as you might guess a list in which elements can only join at the end and leave at the head. So you need to think very clearly about the precise structure involved if you are reading about this kind of processing.

There is one final idea associated with lists which we introduce now because it will be used in later chapters. It may already have struck you that, while the variable length characteristic of a list makes it easy to append items (or other lists) to the end of an existing list, or to take the 'head' item from a list, slotting items into the middle of the list is not nearly so convenient. However, since the order of the items in the list is a significant feature, it is often necessary to do just that. To take a simple case, if British Rail were to introduce into the ordered list of stations ⟨Newport, Cardiff, Bridgend, Neath, Swansea⟩ a new stop *between* Bridgend and Neath, it would be necessary to rewrite the whole list in order to accommodate this in its correct position.

The *linked list* is a device which, among other advantages, helps to overcome this problem. Each item on the list is given an *address*, the word being interpreted very much as it is colloquially — a piece of information enabling the location of the item to be found. In its simplest form this address can consist merely of a numbering of the items on the list:

1. Newport
2. Cardiff
3. Bridgend
4. Neath
5. Swansea

Then we add to each item on the list (or *record*, to use a data processing term) a *pointer* which tells us where to go next. Again at its most basic, this could give us:

Address	Item	Pointer
1	Newport	2
2	Cardiff	3
3	Bridgend	4
4	Neath	5
5	Swansea	0

Here a pointer to 0 means we have come to the end of the list. We also need to know where to start; here 'start=1' would do. Then all the items on the list are linked in order by their respective pointers — hence the name.

You may be thinking that this is all both long-winded and obvious. But the usefulness of the method becomes apparent if we wish to insert the extra station Port Talbot in its proper place between Bridgend and Neath. Instead of rearranging all the existing information, with consequent alterations to addresses and pointers, we can do as follows:

Start=1

Address	Item	Pointer
1	Newport	2
2	Cardiff	3
3	Bridgend	6
4	Neath	5
5	Swansea	0
6	Port Talbot	4

This only requires the addition of the extra item at the end of the other data, and the alteration of a single existing pointer — the one attached to Bridgend. Of course, we could apply a similar idea to the addition of any number of items at various points in the list.

This type of linked list is said to have *forward pointers*. More complex structures involve both forward ('where to go next') and backward ('where we have just come from') pointers, not to mention circular lists. However, this is not the place for a deeper discussion of the topic; if you are interested, you can pursue it in books such as Kolman and Busby (see 'Suggestions for further reading').

Meanwhile, here are a few exercises to test your understanding of lists, followed by some more general problems covering the whole material of Chapter 2.

Exercise 2.10.1
The characters of the word COMPUTE are stored as a list L1 thus: $\langle C, O, M, P, U, T, E \rangle$. Find tail (tail(L1)) and head(tail (L1)).

Exercise 2.10.2
List L2 = $\langle D, I, S \rangle$. Find append (L2, Tail(tail(tail L1))).

Exercise 2.10.3
Re-write the linked list below to include Tuesday and Thursday in their proper place in the sequence.

Start=1

Address	Day name	Pointer
1	Sunday	2
2	Monday	3
3	Wednesday	4
4	Friday	5
5	Saturday	0

PROBLEMS

1. Students in the final year of a computing course may elect to take one of three optional subjects: Advanced Object-Oriented Programming (subject A11), Introductory Object-Oriented Programming (subject I11), and Networking (subject A12). The prerequisites for these subjects are as follows.

   To take A11, a student must have taken courses I5 (Computing Languages) and I8 (Structured Programming), and achieved credit level marks in at least one of these, and a pass in the other.

   To take I11, the same combination is required, but pass level marks in both subjects are sufficient.

   To take A12, a student must have passed I5 and also passed either I4 (Systems Design) or I2 (Introduction to Information Systems) with credit.

   By defining suitable sets of students, express these conditions in terms of set operations. Illustrate your solution by sketching Venn diagrams.

2. A list of customer codes takes the form ⟨P01, P02, B03, P04, B05, B06, B07, P08⟩. The prefix denotes whether the customer is a private (P) or business (B) customer. Write an algorithm (either in the form of an action diagram or other structured format) using the list operations 'head', 'tail' and 'append', to read this list and select all private customers, who are to be stored in numerical order in a separate list.

3. Using set and n-tuple notation, devise a means of representing the structure shown in Fig. 2.12, which illustrates the movements of raw materials between depots and manufacturing

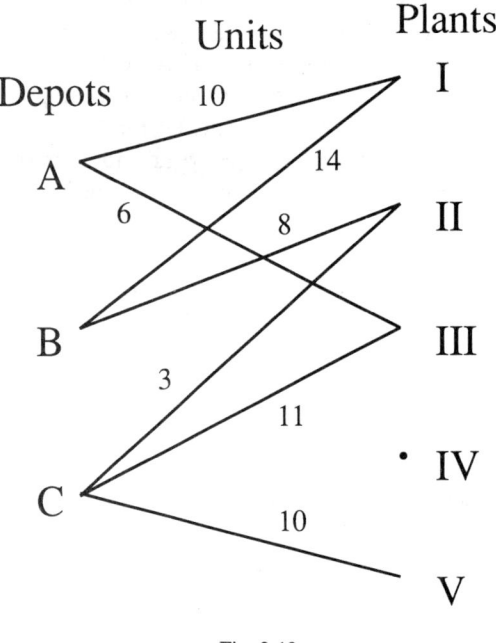

Fig. 2.12.

plants. How could your representation be used to identify all plants being supplied only by depot C?

3
Elementary logic

3.1 INTRODUCTION

This chapter covers two areas which can be included under the title of 'logic'. First we look at the algebra of 'logical' or 'Boolean' quantities — quantities which, instead of taking on numerical values as algebraic variables do, take on the values 'true' or 'false'. Such quantities are, as you may already have discovered, very useful in programming, enabling us for example to set error flags which have the value 'true' at the end of the run if no errors have been found, and 'false' if an error has appeared. This area of logic has many close analogies with set theory, and so we will not need to go into quite so much detail as we did in the last chapter.

Secondly, and leading on naturally from the first part of the chapter, we look at methods of proving results true or false: what can, and what cannot, logically be deduced from a given premise? This is a very important question not only when attempting to construct proofs of mathematical results, but also in the field of program proving: does this algorithm do what it is supposed to do? will the program work for all values of all the parameters? and so on.

3.2 STATEMENTS

We start off with some terminology.

The building blocks of logic are *statements* — simple declarative sentences which must be either true or false, but cannot be both at once. This is analogous to the situation we had in the last chapter, where an element must either belong to a given set or not. As we mentioned at that point, there is currently much interest in 'fuzzy' versions of these concepts, and 'fuzzy logic' is proving to be extremely useful in computing and other applications, but as with fuzzy set theory, the concepts are far more difficult that those of ordinary true/false logic.

Your work in computing may already have introduced you to the idea of a statement, under the name of a *logical* or *Boolean* quantity, so called after one of the founders of mathematical logic, George Boole. Just as an algebraic quantity takes on numerical values, so a Boolean quantity takes on Boolean values. However, unlike algebraic values these are limited in number to two — 'true' or 'false', often abbreviated to T and F. In set-theoretic terms, the value of a Boolean quantity must belong to $\{T, F\}$.

Thus, 'John is a bank manager' is a statement, which will be true if John *is* a bank manager, false if he is not. Likewise, 'The Earth is flat' is a statement, which happens not to be true; 'This sentence is a statement' is a true statement, but 'If it rains tomorrow and you go out you will get wet' is not a statement, since it is not a simple declarative sentence, even through it is clearly true.

Problems can arise with sentences such as 'Stanley is talented'; this is certainly a simple declarative sentence, but whether it is true or not is a matter of value-judgement! We will try to avoid such cases in our examples, and advise you to do likewise. It is for this reason that logicians often use nonsense-statements such as 'all widgets are wodgits' in an effort to avoid any overtones from everyday language; however, this makes the subject seem unnecessarily abstract, so we will try to steer clear of such examples also.

3.3 PREDICATES

The examples of statements given in Section 3.2 were all, so to speak, 'constant statements': having a fixed value of 'true' or false', just as a constant in algebra has a fixed numerical value. Any logical expression whose value is always constant — either true or false — is called a *tautology*.

However, particularly in programming, we often want to use statements which have a variable value, in which case the idea of a *predicate* is useful. This is the name we give to an expression containing a variable which becomes a valid statement, and therefore takes a value of 'true' or 'false', when the variable is given a specific value.

An example will make this idea clearer. Suppose that $P(x)$ is the proposition 'x is a perfect square'. As it stands, $P(x)$ is not a statement, since it is not possible to determine whether it is true or not. But if we let $x=9$, then $P(9)$ takes the value 'true'; similarly $P(5)$ is false, and so on.

You can see that the values which the predicate can take are, as with any statement, the Boolean quantities 'true' and 'false' or T and F. It is also clear that the predicate is a function of x. Putting these two ideas together we often refer to predicates as Boolean-valued functions; they relate the set of possible values of x to the set $\{T, F\}$.

Exercise 3.3.1
Evaluate $P(y) = $ 'y is a positive integer greater than 1 but less than 10' for each y in the set $\{0, 1, 2, 4, 8, 16\}$.

There are a couple of pieces of notation which can save space in defining statements and predicates. We often want to make a statement such as 'x^2 is a non-negative number for all real values of x', which is rather lengthy. It can be abbreviated by the use of the symbol ∀, an upside-down A, which is read as 'for all'. Thus the statement in question could be written as

$$\forall x \in \mathbb{R}, \quad x^2 \geq 0.$$

Sometimes rather than saying that a statement holds for *all* values of a quantity, we want instead to say that there exists at least one value for which it holds. Thus we can say that for any positive real number x, there exists a bigger real number. The shorthand for 'there exists' is ∃, an E written backwards, and the statement in question can be written using ∃ and ∀ thus:

$$\forall x \in \mathbb{R}, \exists y \in \mathbb{R} : y > x.$$

(Remember that the : stands for 'such that'.)

We are not suggesting that you make heavy use of these symbols; they are, after all, only a way of saving words, so if you find it easier to write out statements and predicates in full, by all means do so. However, you should be able to cope if you find the notations used in other books, and occasionally later in this one.

3.4 COMBINING STATEMENTS

Having defined statements, which are the 'mathematical objects' relevant to logic, we next need some operations for combining them. There are three such operations:

$\bar{a}$ means 'not a' or 'the opposite of a'.

$a \wedge b$ means 'a and b'.

$a \vee b$ means 'a or b or possibly both'. (This is sometimes known as the *inclusive or*, and is not the way 'or' is interpreted in colloquial English; when I say 'Have an egg or a cheese sandwich', I do not usually expect you to take both! The colloquial usage is the *exclusive or*, which is also used in some computing languages — so take care.)

These logical operations may remind you of the operations of set theory we introduced in Chapter 2. An element belongs to $A \cup B$ if it is in A or B, or possibly in both — the same form of words which we used to define the logical operator ∨ in the last paragraph. There is thus a connection — at present we put it no more strongly than that — between union of sets and the 'or' operator in logic.

Likewise, an element is in the set $A \cap B$ if it is in both A *and* B; so intersection relates to the logical 'and'.

Finally, an element is in A' if it is *not* in A; so the set complement corresponds to the logical negation. We will see in a later section just why these correspondences exist.

The compound statements resulting from the use of logical operators can be true or false just as simple statements can be; the truth or falsehood of the compound statement will depend on the truth or otherwise of the simple statements being combined. For example, if $a=$ 'This book is in English', $b=$ 'This book is in French' and $c=$ 'This book is about mathematics', then a and c are true while b is false. So $a \wedge c$ is true; the truth of a compound statement involving 'and' requires *both* its component simple statements to be true. Hence $b \wedge c$ is false.

However, $b \vee c$ is true, because for a compound statement with 'or' to be true we require only one *or* the other simple statement to be true — not necessarily both. Of course, if both *are* true the compound statement is still true — $a \vee c$, for example, is true. Only if *both* simple statements were false would the 'or' statement be false also.

The truth or falsity of a negation is easily determined, being the opposite of that for the statement being negated. Since statement b above is false, $\bar{b}$ will be true; since a is true, $\bar{a}$ will be false.

To explain all this in words is, as you can see, a lengthy business, but what we have said above can be summarized quite easily in the form of a *truth table*. Just as an algebraic expression may be evaluated by giving definite values to the variables involved, so a Boolean expression can be evaluated by giving Boolean (true/false) values to the simple statements involved. A truth table is a way of showing the value of the compound expression for all possible combinations of truth/falsity in the component simple statements. This is not such a tall order as might at first appear; since there are only two possible Boolean values, there will only be four possible combinations in a case where two simple statements are being combined. (The calculation of this 'number of possible combinations' is an example of a process to which we shall be returning in Chapter 9.)

The truth table below shows the idea for $a \wedge b$.

a	b	$a \wedge b$
T	T	T
T	F	F
F	T	F
F	F	F

Exercise 3.4.1
Set up the truth table for $a \vee b$. (The compound statement should be false *only* in the case where both a and b are false).

The table for $\bar{a}$ is even simpler.

a	$\bar{a}$
T	F
F	T

The correspondences with set theory pointed out above should have caused you

Sec. 3.4]   Combining statements   51

to wonder what, in logic, might play the part of the identity elements for the operations $\vee$ and $\wedge$. In the case of the $\wedge$ identity, for example, we are looking for a statement — call it X — which is such that $A \wedge X = A$ for all statements A. Thus if A is true then $A \wedge X$ must be true also, and if A is false so must $A \wedge X$ be. If you look back at the truth table for $\wedge$ you will see that any statement satisfying this requirement must *always* have the value 'true'. In other words, a statement which is tautologically true acts as the identity element for $\wedge$, and corresponds to the universal set in set theory; we might think of it as a statement which is universally true.

Exercise 3.4.2
Determine in the same way the identity element for the operation $\vee$ — a statement X such that $A \vee X = T$ if $A = T$ and $A \vee X = F$ if $A = F$. (You should find, not too surprisingly, that the answer is a statement which is tautologically false.)

Naturally we are not limited to combinations of only two simple statements. More complex expressions can be a powerful tool in programming for 'filtering out' particular values, or ranges of values, of variables. We give a few examples, both concrete and abstract, followed by a further exercise for you to try yourself.

Example 1
Evaluate the expression 'x is an even integer and x is divisible by 3 or x is bigger than 5' for $x=10$, $x=9$, $x=6$.

Answer
When $x=10$, the statement is true (x is both even and >5). $x=9$ makes it false, because 9 is not even, while for $x=6$ all three properties are satisfied, hence the statement is true.

Example 2
Let $P(x) = $ 'x is a perfect square', and let $Q(y) = $ 'y is divisible by 3'. Evaluate $P(x) \wedge Q(y)$ for $x=12$, $y=9$, and for $x=9$, $y=12$.

Answer
When $x=12$, $y=9$, $P(x)=F$ and $Q(y)=T$. So $P(x) \wedge Q(y)=F$. When $x=9$, $y=12$, $P(x)=T$. So $P(x) \wedge Q(y)=T$.

Exercise 3.4.3
Evaluate $P(x) \vee Q(y)$ for the predicates defined in Example 2 above, and the given values of x and y.

Exercise 3.4.4
Let $P(x) = $ 'x is a female', and $Q(x) = $ 'x is British' (note that here we are using the *same* value of x in both predicates). Assign a Boolean value to the statements $P(x) \vee Q(x)$, $P(x) \wedge Q(x)$, when x takes the values:

52   Elementary logic   [Ch. 3

Queen Elizabeth I, Charles Babbage (if you do not know who he was, you should!), Joan of Arc, Albert Einstein.

3.5 TWO IMPORTANT LOGICAL IDENTITIES

The analogy between set theory and logic tentatively suggested in the last section enables us to deduce the form of two very important laws. Suppose we want to investigate the logical expression $\overline{(A \vee B)}$, to find out what happens if we 'take the brackets away'; by now we hope you are too wary to fall into the trap of assuming that the 'obvious' answer $\overline{A} \vee \overline{B}$ is necessarily correct.

Instead, let's see if looking at a set-theoretic analogy in terms of a Venn diagram can shed any light on the question. What is another way of expressing $(x \cup y)'$? Fig. 3.1 suggests that it is equivalent to $x' \cap y'$. So perhaps $\overline{(A \vee B)} = \overline{A} \wedge \overline{B}$.

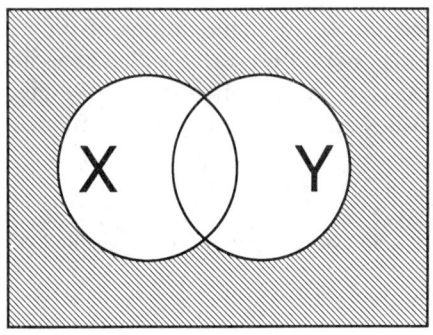

Fig. 3.1.

This does not, of course, constitute a proof, but having decided what form the identity might take, we can verify it by means of a truth table. If the expressions $\overline{(A \vee B)}$ and $\overline{A} \wedge \overline{B}$ have identical truth tables, then their identity is established. It is as well not to try to go directly to the true/false values of the complex expression, but to build it up gradually as shown below.

A	B	$A \vee B$	$\overline{(A \vee B)}$	$\overline{A}$	$\overline{B}$	$\overline{A} \wedge \overline{B}$
T	T	T	F	F	F	F
T	F	T	F	F	T	F
F	T	T	F	T	F	F
F	F	F	T	T	T	T

The identity of the fourth and last columns shows that the relation $\overline{(A \vee B)} = \overline{A} \wedge \overline{B}$ is indeed true; in fact we can go further and write $\overline{(A \vee B)} \equiv \overline{A} \wedge \overline{B}$, the symbol $\equiv$ meaning 'is identically equal to' or 'is equivalent to' — that is, the two sides of the expression are equal under all circumstances, irrespective of the value given to A and B.

Exercise 3.5.1
Show by writing out a truth table that $\overline{(A \wedge B)} = \overline{A} \vee \overline{B}$.

(It can be tricky, particularly for longer tables involving more variables, making sure that you have included all combinations of true/false values. You may like to note that if we equate T with 1 and F with 0, then the sequence of true/false values of A and B in the first two columns of the above table can be interpreted as the sequence of 2-bit binary numbers, written in reverse. Similarly if we had three variables to deal with, we could use the 3-bit binary numbers, and so on. Some books actually use 0/1 rather then T/F, but this can tempt you to carry the analogy with binary arithmetic too far.)

The two laws we have established in this section are of great importance; they are known as De Morgan's laws after the nineteenth-century British mathematician Augustus De Morgan. In order to give them a more than abstract significance, we conclude the section by interpreting them in the context of a practical example.

Suppose $A(x)$ is the predicate 'x is an even number' and $B(x)$ is 'x is divisible by 3'. Then $\overline{(A(x) \vee B(x))}$ means 'x is neither even not divisible by 3', which can also be expressed as 'x is not even AND x is not divisible by 3'. But this last sentence can be written symbolically as $\overline{A(x)} \wedge \overline{B(x)}$; so $\overline{(A(x) \vee B(x))} = \overline{A(x)} \wedge \overline{B(x)}$, just as the first of De Morgan's laws states.

Exercise 3.5.3
Using the same two predicates, interpret the second law.

3.6 THE ALGEBRA OF LOGICAL OPERATIONS

In the last chapter we went to a great deal of trouble to verify that set-theoretic operations satisfied a particular collection of properties which were summed up on p. 37. The same set of properties applies also to Boolean operations, with $\vee$ playing the role that union played in set theory, $\wedge$ that of intersection, a universally true statement acting as universe, and a statement which is always false as the empty set. Any system of operators and operands which satisfies this set of properties is in fact called a *Boolean algebra*.

For logical operations we therefore have the following:

$A \vee B = B \vee A$ and $A \wedge B = B \wedge A$

$A \vee (B \vee C) = (A \vee B) \vee C$ and $A \wedge (B \wedge C) = (A \wedge B) \wedge C$
$A \vee (B \wedge C) = (A \vee B) \wedge (A \vee C)$ and $A \wedge (B \vee C) = (A \wedge B) \vee (A \wedge C)$
$A \vee \overline{A} = T$ (a statement or its negation must always be true) and
$A \wedge \overline{A} = F$ (a statement and its negation cannot simultaneously be true)
$A \vee F = A$ and $A \wedge T = A$.

You should compare this with the properties of sets established on p. 37 and convince yourself that they are essentially the same. It is possible to define general Boolean algebras — other systems of operators having the same set of properties — but in this book you will only be encountering the two examples already discussed. 'Ordinary algebra' is of course NOT a Boolean algebra; as we have already noted it fails because addition is not distributive over multiplication.

3.7 AN APPLICATION OF LOGICAL ALGEBRA

We conclude the first part of this chapter by looking at an example of the use of logical operators to select certain subsets from a universal set. You will see more examples like this in Section 8.4 on relational databases.

Suppose that a small firm specializing in direct-mail sales keeps a computerized file of all its customers. For each customer there is a record indicating whether the customer belongs to each of the following sets:

- A. Regular customer
- B. Prompt payer
- C. Bad debtor
- D. Overseas customer
- E. New customer (first transaction within the last 12 months)
- F. Account customer

Of course, these categories overlap, so that a single customer could be both a new customer and a prompt payer. We will write $A(x)$ to denote the predicate 'x is a regular customer', and similarly for the other categories. Thus $A(\text{Smith}) = T$ if Smith is a regular customer, $A(\text{Stevens}) = F$ if Stevens is not.

Now imagine that the firm wishes to compile a list of all new overseas customers who are not bad debtors. It could do this by means of the compound expression $E(x) \wedge D(x) \wedge \overline{C(x)}$. This expression would be evaluated for each customer on the file in turn, and only those for which it has the value T will be added to the list.

Likewise customers who are either bad debtors or non-account customers, and who are overseas, will be those for whom $[C(x) \vee \overline{F(x)}] \wedge D(x)$ is true. In this way we can pick out groups of customers satisfying complex combinations of criteria in a very simple way.

Exercise 3.7.1
Write down logical expressions which would be true for (a) customers who are neither account

nor regular customers, but are prompt payers; (b) customers who are overseas, and are either regular or account customers, but not both.

3.8 IMPLICATION

The logical operations which we have discussed so far are not what most people would understand by 'logical' in the colloquial sense. If you say to someone, 'That's not logical', what you generally mean is, 'It doesn't follow' — your conclusion, in other words, does not follow from your *premise*, or initial statement, in a logical fashion. In the rest of this chapter we will examine what *can*, and what cannot, be logically concluded from a certain premise, and then apply these ideas to the question of how one can construct logically valid proofs.

The formal notation which we use to mean that one statement 'follows' from another is the symbol $\Rightarrow$; the notation $A \Rightarrow B$ is read as 'A implies B'. Another helpful way of interpreting this is 'B follows by logically correct argument from A'. The implication sign may disguise quite a long series of steps in a proof, as we shall see.

Implication is really just another logical operator like $\vee$ and $\wedge$, and the truth or otherwise of the compound statement $A \Rightarrow B$ for various values of A and B can, as for statements involving those operators, be shown by means of a truth table. In this case it is probably easiest to give the table first and then investigate what it means.

A	B	$A \Rightarrow B$
T	T	T
T	F	F
F	T	T
F	F	T

Let's examine each row of this table in turn. The first row tells us it is true that a true conclusion follows from a true premise. This is reassuring, since arguing by correct implications from true premises is the foundation of most mathematical proofs!

The second row says that if we get a false conclusion from a true premise, the implication must be false — in other words, there must be something wrong with the logic of the argument. This is in line with our intuitive feelings; if you could start from a true premise, proceed by entirely correct steps and suddenly find yourself with a wrong or meaningless conclusion, you would feel very insecure.

It is the remaining two rows of the table which often cause students encountering them for the first time to feel puzzled. The third row states that we can start from a false premise, proceed by logically quite correct arguments, and arrive at a true conclusion. It is entertaining to try constructing illustrations of this; for example, starting from the statement 2=4, we can say that 4=2, hence addition gives 6=6, which is certainly true (this actually works with any pair of numbers). There is nothing wrong with the argument here — it is the initial statement which is false. So a true argument has led from a false premise to a true conclusion.

The last row of the table means that a true argument can lead from a false premise to a false conclusion. To take a non-mathematical example, we could make the statement 'All cows have three legs', and then argue, 'This animal is a cow. Therefore it has three legs'. Again there is no flaw in the logic of the deduction — we have reached a silly conclusion because the premise was silly.

In fact, the last two rows of the table are simply telling us that if you start off with a false premise then, however, logically you argue, there is no knowing where you may end up.

We can now incorporate implication into our evaluation of compound logical expressions. For instance, if we wish to evaluate the expression.

$$(A \lor B) \land C \Rightarrow D$$

when $A=$T, $B=$T, $C=$F and $D=$T, we can substitute for A, B, C and D exactly as we would in evaluating a numerical expression, to get (T$\lor$T)$\land$F$\Rightarrow$T. Now T$\lor$T=T, reducing the expression to

$$T \land F \Rightarrow T.$$

But T$\land$F=F, so we can make the further reduction to F$\Rightarrow$T. Finally, looking at the third row of the truth table for implication tells us that F$\Rightarrow$T has the value T. We have therefore shown that

$$(T \lor T) \land F \Rightarrow T \text{ simplifies to } T.$$

Be careful not to omit the last step in this evaluation; many newcomers to the topic get correctly as far as F$\Rightarrow$T but then fail to carry out the last simplification. Remember that if you are evaluating a logical expression the final result should be a single value — T or F — just as in evaluating an arithmetic expression, the final result is a single number.

This section has been fairly abstract so far; we conclude with a practical example which should give you a better grasp of the truth table for implication. Consider the proverb 'There is no smoke without fire'. This can be translated into a logical implication by defining A to be the statement 'There is smoke', and B the statement 'There is fire'. The proverb then becomes $A \Rightarrow B$, which corresponds when there *is* smoke and fire to the first row of the truth table, T$\Rightarrow$T, which is true.

The second row would suggest that there is smoke, but no fire — T$\Rightarrow$F. This is patently not possible, so the value it takes is F.

The third row, F$\Rightarrow$T, translates into 'there is no smoke but there could be fire', which can indeed be true — the fire is banked down so that no smoke appears. Thus F$\Rightarrow$T takes the value T.

Finally, the last row F$\Rightarrow$F tells us that there is no smoke and there could be no fire, which again can be true.

3.9 NECESSARY AND SUFFICIENT CONDITIONS

If a statement A is linked by a true implication to another statement B, we say that A is a *sufficient condition* for B. This may sound like more jargon, but it actually agrees with the ordinary everyday meaning of 'sufficient', as a couple of examples will show.

Example 1 (an arithmetical example):
Let $p(x)$ be the statement 'x is a positive integer divisible by 6', and $q(x)$ be 'x is a positive integer divisible by 3'. We can certainly say $p(x) \Rightarrow q(x)$, and clearly for a number to be divisible by 3 it is sufficient that it should be divisible by 6 — sufficient meaning 'enough', so that no further conditions are required. $p(x)$ is therefore sufficient for $q(x)$, in line with our definition above.

Example 2 (a more practical example)
Let $A(x) = $ 'company x makes a loss three years in succession', and $B(x) = $ 'company x goes bankrupt'. Suppose that for a particular company N we know that $A(N) \Rightarrow B(N)$ — the company will go bankrupt if it makes a loss for three successive years. Then again $A(N)$ is sufficient for $B(N)$; three successive losses are sufficient, or enough, to ensure bankruptcy, no additional conditions being needed.

Notice that in general implication is not reversible — just because $A \Rightarrow B$ it isn't necessarily true that $B \Rightarrow A$. In the last example, for instance, the fact that the company has gone bankrupt does not imply that it made losses three years running. That *might* be the case, but there could be other reasons for the bankruptcy. So $B(N)$ is not sufficient to enable us to deduce the truth of $A(N)$.

This is a point which beginners often fail to grasp when asked to prove a result. If you have studied geometry you may recall that it can be proved that an isosceles triangle (one with two equal sides) has two equal angles. Logically this can be expressed by saying

triangle isosceles $\Rightarrow$ triangle has two equal angles.

If you are asked to prove this, starting off by *assuming* that the triangle has two equal angles and proving that it is isosceles *will not do*! This is equivalent to proving

triangle has two equal angles $\Rightarrow$ triangle isoceles,

and the implication may not be reversible.

So if $A \Rightarrow B$ means that A is sufficient for B, and this is not in general reversible, what *can* we say about the relation of B to A? We say that B is *necessary* for A, the usage again being in line with ordinary speech. 'B necessary for A' means A can't be true unless B is also true — though other conditions may be required as well to *ensure* that A is true. In Example 1 above, $q(x)$ is necessary for $p(x)$ — for a number to be divisible by 6, it is necessary for it to be divisible by 3. It isn't sufficient — the condition of being an even number is also necessary to *ensure* that x is divisible by 6 — but certainly x cannot be divisible by 6 unless the condition of divisibility by 3 is present.

Again, in Example 2, the company having gone bankrupt is a necessary condition for it to have made a loss in three successive years — if it hasn't gone bankrupt, then it can't have made three successive losses; A can't be true unless B is true, so B is necessary for A.

The expression $A \Rightarrow B$ is sometimes read as 'if A then B', which you may find easier when thinking about necessary/sufficient conditions. 'If A then (always) B' indicates that A is sufficient for B. To decide whether a condition B is necessary for A, perhaps the best way is to ask, 'Is it true that if *not* B then not A either?' — in other words, we cannot have A unless B is true? If the answer to this question is yes, then B is necessary for A.

There are cases where the implication is reversible, and we can say both $A \Rightarrow B$ and $B \Rightarrow A$. This is written as $A \Leftrightarrow B$, and in this case A is both a necessary *and* a sufficient condition for B (and of course vice versa also). An example of this would be the statements $E(x)$, meaning 'x is an even number', and $T(x)$, meaning 'x is divisible by 2'. These can be linked by the true implication $E(x) \Leftrightarrow T(x)$. Here we can say that for a number to be even it is both necessary (we can't manage without it) and sufficient (nothing else is needed) for it to be divisible by 2. This sort of statement has an elegant self-containment about it — there are neither unnecessary extras dangling around, nor are we short of anything needed.

The double-ended implication is often read as 'B if and only if A'; B is true if A is true, and *only* if A is true — not otherwise. The phrase 'if and only if' may be abbreviated to 'iff', and you will encounter the phrase 'an iff condition' as shorthand for 'a necessary and sufficient condition'.

Exercise 3.9.1
In order to gain her qualification, a student must pass the examination, and either gain over 40% in all course assessments, or gain over 45% in all except one. Using the following notation:

> P(student) = 'student passes exam'
> Q(student) = 'student gets over 40% in all assessments'
> R(student) = 'student gets over 45% in all except one'
> S(student) = 'student gets qualification',

write down all the true implications which link P, Q, R and S, and interpret them in terms of necessary and sufficient conditions.

Exercise 3.9.2
Consider the following series of predicates, in which x stands for a positive integer:

> $H(x) =$ 'x is a power of 3'
> $N(x) =$ 'x is a power of 9'
> $E(x) =$ 'x is an even power of 3'
> $O(x) =$ 'x is an odd number'

Complete the table overleaf by indicating in each case whether the implication is true or false. For example, one entry has been completed to show that $H \Rightarrow N$ is false.

⇒	H	N	E	O
H		F		
N				
E				
O				

Interpret in terms of necessary and sufficient conditions. Are there any pairs of statements which are both necessary and sufficient for each other?

Exercise 3.9.3
Which of the following are true?

(a) A rectangle is a square if and only if its adjacent sides are equal.
(b) If an electric light works, then the power supply must be on.
(c) If the power supply is on, then an electric light must work.

3.10 METHODS OF PROOF

In the last section we explored some of the fundamental ideas behind the process of logical argument. Now we apply those ideas to arrive at different methods by which a sound mathematical proof may be constructed.

It is important that you realize, in reading this section, that it is not the individual results being proved which are of interest, but rather the general processes of which the particular proofs we give are merely examples. For this reason, we have chosen to present proofs of very simple results in most cases, so that you will not have the problem of being unable to see the wood for the trees. If there are, in spite of this, details of algebra or arithmetic involved in these proofs which you do not follow, try not to worry but concentrate on the underlying logic.

(a) Constructive proof

The name given to our first method reflects the idea that we can prove something is possible by doing it, or at least providing a method by which it can be done. At a trivial level, if I say to you, 'I can swim the Channel', and you reply, 'Prove it!', I am unlikely to sit down and offer a logical argument to demonstrate the truth of my claim — and if I did, you would probably not be convinced. What I would have to do is to swim the Channel, thereby offering conclusive evidence that my statement was true.

Such a direct approach can work well in a scientific context too. Many proofs in geometry are of this kind: for example, to show that a circle can be drawn through any three points not in a straight line, we actually demonstrate how the circle can be constructed. Often, too, such proofs take the form of giving an algorithm by which the required process can be carried out. For example, if you wish to prove the truth of the statement that 'any odd number can be written in the form $2k+1$, for some value of n', the very simple algorithm for expressing an odd number in this form would be given by the action diagram below.

```
┌─Read number n
│ ┌─If n odd
│ │    p = n−1
│ │    k = p/2
│ │    n = 2k+1
│ └─endif
└─end
```

Unfortunately, it is by no means easy to find such constructive proofs in general; mathematics abounds in what are called 'existence proofs', which show that something exists without offering a way of producing the thing in question. In the field of coding and information theory, for instance, about which you will be reading in Chapter 10, certain classes of very elegant and efficient code can be proved to exist, but there are no constructive methods for actually setting up the coding/decoding algorithms. The codes which *have* been constructed have often been arrived at in a roundabout way, and do not approach in efficiency those which mathematics tells us must exist — somewhere!

(b) Direct proof

This is the category which includes those proofs beginning with a true premise and simply proceeding, by a longer or shorter series of true implications, until the truth of the statement we set out to prove is reached. The validity of this type of proof is a consequence of the first row of the truth table for implication on p. 55.

As an illustration of this method, suppose we wish to show that the square of an odd number is always of the form $4p+1$, for some integer p. Then we could proceed thus.

Let n be an odd number. The truth of the fact that n is odd is our initial true statement or *premise*.

n can be written as $2k+1$ for some integer value of k — this is a true implication, since all odd numbers can be put into this form (see the action diagram in section (a) above).

$$n^2 = (2k+1)^2$$

— squaring both sides.

$$n^2 = 4k^2 + 4k + 1$$

— this is true by elementary algebra, but don't worry if you could not work it out for yourself.

$$n^2 = 4(k^2+k) + 1$$

— taking out a common factor of 4 from the first two terms.

$$n^2 = 4p + 1$$

— writing p for k^2+k.

Thus n has the form which we set out to prove.

There are often steps in this kind of proof which look surprising to the novice; maybe the step of writing an odd number as $2k+1$ was a bit surprising to you. There are two things to be said about this. First, there is a sort of repertoire of 'tricks' which one begins to acquire after doing a certain amount of this kind of thing; writing odd numbers as $2k+1$ and even ones as $2k$ would be an example of this kind of 'trick'.

Secondly, it is perhaps unfortunate that proofs in textbooks are always presented in what one might call 'finished form' — which is usually anything but the way in which they were originally reached. We are sure that Pythagoras did not suddenly say to his friends over dinner, 'By the way, the square on the hypotenuse is equal to the sum of the squares on the other two sides, and to prove it you do this construction ...'. He had probably noticed that the squares *looked* as if they might have such a relationship to one another, and then spent some time working out if really this was true, and if so, how it could be proved, with scribbled calculations and a few trips down blind alleys. Many important results have been reached by first arguing from the particular to the general and conjecturing the form of the result, then proving in the proper mathematical sense that the conjecture 'works' — more of this a little later.

In fact, the result we have just proved — that the square of an odd number is a multiple of four plus one — becomes 'obvious' if you start looking at the squares of specific odd numbers — $3^2 = 9 = (4 \times 2) + 1$, $5^2 = 25 = (4 \times 6) + 1$, and so on. But this is not a *proof* — which is why the argument above was needed.

So do not get discouraged if your efforts to prove results do not look as tidy, and make such a beeline from premise to conclusion, as those in textbooks — you are following a fine tradition.

(c) Proof by contradiction

This is a method based on the fact, implicit in the last row of the truth table for implication, that if a true argument leads you to a false conclusion, then the initial premise must have been false also. The method is particularly useful if we want to prove results of a negative kind, which say that something is *not* the case.

We give two illustrations of this type of proof. The first is exceedingly simple; the second, while less so, is something of a classic example.

Example 1
A fraction of the form p/q, where p and q are both even, cannot be in its lowest terms.

Proof
We begin, as always with proofs of this kind, by assuming that the thing we are trying to prove false is in fact true. In symbolic terms, if we want to prove that $A=T$, we assume that actually $A=F$, or equivalently that $\bar{A}=T$. Here we therefore assume that there *is* a fraction, in its lowest terms, for which p and q are both even. 'In its lowest terms' means that p and q have no common factor, and therefore the fraction cannot be cancelled.

However, if p and q are both even, then we can write $p=2m$ and $q=2n$. The fraction thus becomes $2m/2n$, which has a common factor of 2 top and bottom, and *can* therefore be cancelled. It is thus not in its lowest terms, contradicting the original statement. We have arrived at a contradiction, and the only way this can have arisen, since all the steps of the argument have been sound, is that the original premise was false. So there is *no* fraction in its lowest terms for which p and q are both even — and this is exactly what we wanted to prove.

Example 2
To prove that $\sqrt{2}$ is not a rational number — that is, it cannot be written as p/q where p and q are integers.

Proof
To prove this, start as before by assuming the converse is true, so that we can say $\sqrt{2}=p/q$ for some integers p and q. Furthermore, we can assume *without loss of generality* (that is, without restricting the problem in any way) that the fraction is in its lowest terms. If not, it could be cancelled, and the proof applied to the cancelled fraction.

Squaring both sides gives $2=p^2/q^2$, so that $2q^2=p^2$. Thus p^2 is even, which means that p must be even also (the square root of an even number is even).

Now if p is even, and p/q is in its lowest terms, it follows from the previous example that q cannot be even, but must be odd.

We can now use the fact that p is even to write $p = 2k$. Then $\sqrt{2} = 2k/q$, so that $2 = 4k^2/q^2$, or $2q^2 = 4k^2$, whence $q^2 = 2k^2$. So q^2, being of the form 2 × integer, must be even, and therefore so must q.

We have now shown that q is both odd and even, which is clearly a contradiction. As all the steps in our argument have been logically valid, the original premise, that $\sqrt{2}$ could be expressed as a fraction p/q, must be false. The only possible conclusion is therefore that $\sqrt{2}$ *cannot* be so expressed. (The name for such a number is an *irrational* number; we will be returning to this idea in Chapter 4.)

(d) Counterexamples

This is perhaps a method of disproof rather than of proof. You have probably heard the saying, 'One swallow doesn't make a summer'. In mathematics we can say that one counterexample — one situation in which a result 'doesn't work' — *does* make that result invalid. To disprove a so-called 'theorem', it is enough to find one situation in which it is not true. This is an important fact to bear in mind when going through the kind of thinking described under (b) above — conjecturing the form of a result and then proving it. In the case of 'all squares of odd numbers have the form $4p+1$' we were able to find a valid proof of the conjecture; but had we found one odd number whose square was *not* of this form, the conjecture would have been invalid.

A trivial example would be the statement 'All prime numbers are odd' (recall that a prime number is a number greater than 1 which has no factors except 1 and itself). Since 2 is a prime number, which is not odd, the statement as it stands is untrue — though in this case it can easily be patched up by amending it to read 'All prime numbers greater than 2 are odd'.

This is the idea behind the non-verbal reasoning tests which some of you may have taken as part of aptitude testing for would-be programmers. The sort of test in question usually consists of a set of four or five diagrams, like the ones shown in Fig. 3.2. You are asked to spot the 'odd one out', but in order to do this, you need first of all to formulate some kind of 'rule' obeyed by all but one of the diagrams. The 'odd one out' is then a counterexample to that rule.

The principle also has important applications in program testing. If a program or piece of code can be said to 'work', it must do so for all possible sets of input parameters (sometimes referred to as all 'input states' in the 'input state set'). If you

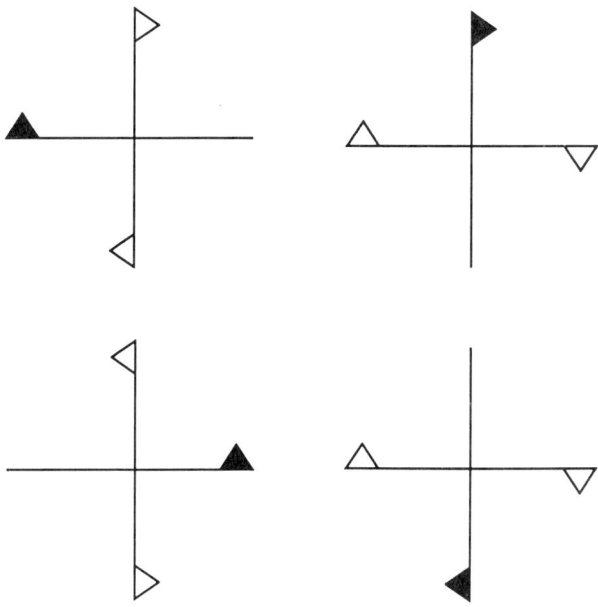

Fig. 3.2.

can find even one member of this set for which an error arises, then the code 'does not work', and you need either to exclude such input states from the allowable input state set, or else modify the code to cope with them. We have often had the experience, in marking students' programming assignments, of crashing a program by means of inputting, say, a fractional value for a parameter when the author of the program had allowed only for the possibility of integer inputs. While this is sometimes viewed as 'unfair', it is a perfectly legitimate way of 'proving' the viability of a program. Some of the difficulties now arising in software testing occur because the programs being written are of such complexity that it is not possible even to list all elements of the input state set, let alone test all possible input states in a reasonable time; then totally different approaches must be used.

However, for the purposes of testing the kind of programs likely to be involved in student assignments, the counterexample idea is a useful one to remember. It is usually the 'odd' values of parameters, rather than the 'routine' ones, which give trouble. For example, in the algorithm for working out the slope of a straight line given the co-ordinates of two points on the line, represented by the action diagram below, it is the 'odd case' when $B=A$ which will give a division by zero and cause the algorithm to fail. (What does this mean in geometric terms?)

```
┌─Begin
│ Read co-ordinates of points [A,C], [B,D]
│ Let slope = (D−C)/(B−A)
│ Print slope
└─End
```

(e) Enumerative proofs

'Enumerating' means counting by explicitly listing all the possible cases which may arise; we will have more to say about enumeration as a method of counting in Chapter 9. In the earlier history of mathematics enumerative proofs were not very popular, one reason for this being that they were viewed as a rather heavy-handed, and more elegant — and shorter — methods were sought. The other reason was that, in many cases, enumerating all the possibilities which need to be tested in order to establish a complete proof of a result is simply not practical for a human operator; it would be prohibitively slow and complex.

However, since the advent of mechanized computation, enumerative proof has become a much more practical proposal, and some results which had for a long time defied what we might call 'analytic' proof have succumbed to this approach.

One notable example of this success is the proof of the so-called 'four-colour theorem', which states that four colours are sufficient to colour any map in the plane so that no two adjacent regions have the same colour. Of course, the number of possible maps is infinite, but it can be reduced by mathematical analysis to a large but finite number of types of map which must be considered. A computer has been used to enumerate all these types, and to show that for all of them four colours suffice. Incidentally this is a nice example of the difference between necessary and sufficient conditions — you can easily demonstrate that four colours are necessary for at least some maps simply by drawing them; it was the proof of sufficiency which was intractable. You will encounter map-colouring problems again in Chapter 5 — though we will not ask you to prove the four-colour theorem!

A subset of enumerative proofs consists of those situations where we enumerate, not all possible cases, but all *significantly different* cases which may arise. This is best seen by an example. Suppose we wish to prove that for any positive integer n, n^2+n is an even number. We can do this by considering the two separate cases n odd and n even, which between them cover all the possibilities.

If n is odd, then $n = 2k+1$, so $n^2+n = (2k+1)^2+(2k+1) = (4k^2+4k+1)+(2k+1) = 4k^2+6k+2 = 2(2k^2+3k+1) =$ even.

If n is even, then $n = 2k$, so $n^2+n = 4k^2+2k = 2(2k^2+k) =$ even.

This is a useful approach to remember in situations where there is a significant difference between the situations n odd and n even (as above), n positive and n negative, and so on.

(f) Proof by induction

This has been left until last because it is probably the trickiest type of proof. It quite often happens in computing that we need to find a 'rule' which will generate for us the terms of a series. A simple case would be the need to pick out every 10th person from a list held on file, starting at the 7th person (what is called in statistics a *systematic sample*). We would probably begin by writing down the first few terms — the 7th, 17th, 27th, ... person — and then generalize quite easily to say that the nth member of our sample will be the $[7+10(n-1)]$th person. (Can you see why $n-1$ and not n?)

However, such a generalization from the particular, though easy to do in this case, does *not* constitute a proof, and in fact can lead you badly astray. It is necessary, having 'guessed' at a general result by looking at a few terms, to prove in a sound logical fashion the truth of that result. Proof by induction is designed to provide that logical method in the case where the result we are aiming to prove is dependent on integer values of a variable n (that is, $n = 1, 2, 3, \ldots$), as in the example of the last paragraph.

We begin by explaining the general principles of the method, and then show how it works in a specific example. We first test if our conjectured result is true for $n=1$, the initial value of n. Then we *assume* that it is true for some value of n greater than or equal to 1, such as $n=k$, and based on this assumption, show that the truth of the result for $n=k+1$ — the next value up — must follow logically. Now we can say that as the result *is* true for $n=1$, and as its truth for $n=k+1$ given truth for $n=k$ has also been established, it must be true for $n=1+1=2$, and thus for $n=2+1=3$, and so on for all integer values.

This may become clearer if we express it symbolically. Let $A(n)$ denote the proposition we are trying to prove true for all integer values of n. The inductive process first shows that $A(1)=T$. We then show that $A(k)=T \Rightarrow A(k+1)=T$. From this it is clear that $A(1)=T \Rightarrow A(2)=T$, which in turn implies $A(3)=T$, and so on.

Students coming to this method for the first time often fall into one of two traps. Either they imagine that it is enough to test the statement for $n = 1, 2, 3, \ldots$, and simply say 'and so on', or they dismiss the whole method as 'a cheat' on the grounds that we are assuming the truth of the result for $n=k$, which is surely what we are trying to *prove*? The first of these is the old mistake of generalizing from the particular, whose dangers we have already pointed out: how do you *know* that just because the result 'works' for $n=1$, $n=2$, or even $n=42$, it won't fail to work for $n=327$?

The second error is a little more subtle, but misses the point of the logical argument. We are not saying 'assume it's true' — full stop. We are first showing that *if* it is true for $n=k$, *then* it is true for $k+1$, and afterwards showing that there *is* a value of k for which our assumption is correct — namely, 1.

To sum up, then, the steps in a proof by induction are:

(1) Show that conjectured result is true for $n=1$.
(2) Assume conjectured result is true for $n=k$.
(3) Arguing from this assumption, show that it is also true for $n=k+1$.
(4) Steps (1) and (3) taken together imply that it is true for $n=2$, $n=3$, and all higher integers.

Now for our illustration of the method.

In Chapter 7 you will be reading about *trees* — structures like the ones shown in Fig. 3.3, consisting of a set of points or nodes connected by edges in such a way that there are no unconnected nodes, but neither are there any closed loops made up of edges. As you can see, the structure looks rather like a stylized picture of a real tree. An obvious question to ask is, how many edges will such a tree have for a given number of nodes?

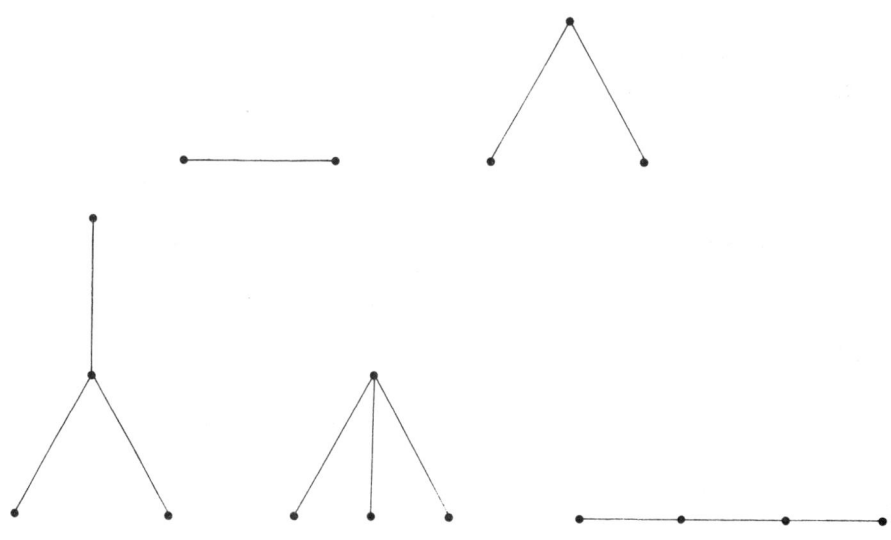

Fig. 3.3.

First, we try to guess at the result by looking at a few specific cases. Fig. 3.3 shows trees with 2, 3, and 4 nodes, and you can see that the numbers of edges are respectively 1, 2, and 3 (even though the 4-node tree can be drawn in three different ways, the number of edges is always the same). So our guess is that if there are n nodes, there will be $n-1$ edges.

Now for the formal proof.

The first step is to test the truth of the result for $n=1$. How many edges has a tree with 1 node got? The answer, of course, is none; when $n=1$, number of edges $= 0 = 1-1 = n-1$. So the conjectured result is certainly true for $n=1$.

Now we assume that the result holds for $n=k$, so that a tree with k nodes will have $k-1$ edges. If this is true, how many will a tree with $k+1$ nodes have? This can be rephrased to ask how many extra edges are created when we add the extra node to the tree? The answer is one. No matter whereabouts in the existing tree structure the extra node is added, no more than one extra edge can be incorporated without creating a closed loop, which is not allowed in a tree; and at least one extra edge is needed otherwise the new node will be unconnected, which is not allowed either. So the introduction of the new node generates precisely one new edge, the new tree therefore having old edges $+1 = (k-1)+1 = k = (k+1)-1$ edges.

Now compare the two statements:

> a tree with k nodes has $k-1$ edges,
> a tree with $k+1$ nodes has $(k+1)-1$ edges.

You can see that the second is of exactly the same form as the first, except that k has been replaced by $k+1$.

Sec. 3.11] When is a proof not a proof? 67

We have therefore established that *if* the result is true for $n=k$, it follows logically that it is also true for $n=k+1$. But we saw that it is certainly true for $n=1$, and so it must be true also for $n=2$, and hence for $n=3, 4, 5$, — indeed, for all positive integer values of n.

If you feel at this point that, even though you accept and understand the method of this kind of proof, you would never be able to construct one for yourself, don't worry — in the problems at the end of the chapter we guide you step-by-step through one such proof, before asking you to attempt one on your own.

3.11 WHEN IS A PROOF NOT A PROOF?

We conclude this chapter with a few cautionary notes as to what does *not* constitute a valid proof, and what pitfalls may lie in wait for those who are a bit casual about their logic. These warnings have been issued at various points earlier in the chapter, but we put them together here, since they cannot be emphasized too much.

First, be very wary of looking at two or three special cases of a result and simply saying, 'Oh well, it seems to work OK'. Suppose you have been asked to find a 'formula' for the sum of the squares of the first n integers. That is, we want to find a way of expressing $1^2+2^2+3^2+ \ldots +n^2$ in terms of n. The first few values of n give the following: $n=1$, $1^2=1$. $n=2$, $1^2+2^2=1+4=5$. $n=3$, $1^2+2^2+3^2=1+4+9=14$. 'Guessing' a result from this is quite a tall order, but we suggest to you that a suitable formula (never mind where we got it from!) is that the sum of the first n squares is $0.5\,(5n^2-7n+4)$. Does this work? How far do you need to go before concluding that it doesn't?

Second, make sure you are not guilty of proving that $A \Rightarrow B$ when you are asked to prove $B \Rightarrow A$. As we saw earlier in the chapter, the two are not in general interchangeable.

Finally, be prepared to test your proofs — or your programs — thoroughly; remember that one counterexample is enough to invalidate a proof, and think clearly through any 'special cases' which may arise, to make sure that they do not cause problems.

PROBLEMS

1. What method of proof do you think might be used in each of the following cases? (Do NOT try to construct the proofs.)

    (a) To prove that a line can be drawn through a given point parallel to a given line.
    (b) To prove that the sum of the first n positive integers is $0.5n\,(n+1)$.
    (c) To prove that the square root of a negative number cannot exist in the ordinary number system.

2. You are going to construct a proof by induction to show that the sum of the first n positive odd numbers is n^2.

    (a) Verify that the suggested result 'works' for $n = 1, 2, 3, 4$. (If you were tackling the problem unaided, this is how you might begin, before 'guessing' at the result on the basis of your four trial calculations.)

(b) Assume that the result is true for $n=k$, that is, that the sum of the first k odd numbers is k^2. Noting that the kth odd number can be written as $2k-1$, we can express this as

$$1+3+5+ \ldots +(2k-1)=k^2 \tag{1}$$

(Why do we need $2k-1$ and not $2k+1$ which we used earlier to represent an odd number?) We now want to establish the truth of the result for $n=k+1$, that is, when the next odd number is included. The next odd number is $2(k+1)-1=2k+1$. Add this to both sides of equation (1) above, to get

$$1+3+5+ \ldots +(2k-1)+(2k+1)=k^2+2k+1 \tag{2}$$

Show that the right-hand side of equation (2) can be re-arranged as $(k+1)^2$.

(c) We have now shown that

if sum of first k odd numbers $= k^2$,
then sum of first $(k+1)$ odd numbers $= (k+1)^2$.

But you have already proved in (a) above that the result is true for $n=1$. We can therefore conclude that it will be true for $n=2$; and if for $n=2$, then for $n=3$, and so on for all positive values of n.

3. A *complete graph* (which you will be encountering again in Chapter 4) consists of a set of *nodes* (or *vertices*) each joined to every other node by a set of *edges*. Fig. 3.4 shows complete graphs with 2, 3, and 4 nodes.

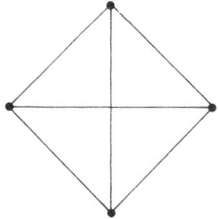

Fig. 3.4.

(a) How many edges does each of these have?
(b) Can you generalize to get a formula for the number of edges in a complete graph with n nodes?

(Hint: try writing a table thus:

Nodes	Edges
1	0
2	1
3	3
4	...).

(c) Check your formula with the solutions on p. 270, and then construct an induction proof (for the correct version of the formula!). (Hint: K_3 has two more edges than K_2 — why?)

4. On a company's computer file of employee names and numbers, each employee's name is followed by a four-digit sequence of 0s and 1s, denoted by x. These are interpreted as follows, with 0=false, 1=true:

   A. The first digit indicates whether the employee is paid monthly.
   B. The second digit indicates whether the employee is female.
   C. The third digit indicates whether the employee has been employed for five years or more.
   D. The fourth digit indicates whether the employee is part-time.

   All employees not paid monthly are paid weekly, and all employees who are not part-time are full-time.
   Thus 0110 indicates a female, full-time, weekly-paid worker who has been employed for five years or more.
   Capital letters are used to denote the corresponding predicates, so that $C(x)$=T if the employee with number x has been employed for 5 years or more. e.g. $C(0110)$=T, $D(0110)$=F, etc.

   (a) Write down logical combinations of these predicates which will select (i) all monthly-paid female workers; (ii) all part-time workers who are monthly-paid males; (ii) all *except* weekly-paid part-time workers.
   (b) Evaluate the expression $D(x) \wedge [B(x) \vee C(x)]$ for x=0000; x=1111; x=1010.

5. (a) Write out a truth table for the expression $\overline{(A \vee B)} \Rightarrow C$. (Hint: the table will have eight rows.)
   (b) Interpret this expression in practical terms if A = 'there is a bug in program X', B = 'computer Y is not functioning properly', and C = 'program X can be run successfully on computer Y'. Can the direction of the implication be reversed?

4

Data representation and manipulation

4.1 INFORMATION AND HOW WE REPRESENT IT

This book is about mathematical ideas needed by people who work with computers, and if you try to describe what computers do in the most basic and general terms, 'handling information' is not a bad start. Whether a computer is carrying out arithmetical calculations, updating a file of customer names and addresses, or controlling the operation of a chemical plant, some kind of processing of information is going on. In the case of the arithmetical calculation, the information is probably in the form of numbers; with the customer names and addresses, most of it will be alphabetic characters; with the chemical plant, all kinds of information could be in use — both quantitative, such as temperatures of furnaces, and qualitative, such as 'on' or 'off' for a particular machine.

If we want to be able to make the computer handle all these different types of information in an effective way, we first of all need to have a clear idea how we ourselves handle information — the means we use to represent it, the laws we take for granted in operating with it, and so on. This chapter is therefore devoted to looking at ways of representing and manipulating information, or data to use a more concise term (we will not involve ourselves here in the technicalities of the distinction between information and data).

The information we deal with will generally be of a quantitative or numerical character. This is not as restrictive as it may seem; frequently the easiest way to deal with non-numeric data such as 'on/off' is to represent it by a numerical code such as 0/1.

We will begin by trying to distinguish between data itself and the mode of representation used for that data, and then go on to reflect on the representation of numerical data, and how the method of representation affects the way we operate with it. This will require you to re-assess the way you think about 'ordinary arithmetic', and perhaps to discover that some apparently unchangeable laws which you have taken for granted for a very long time are actually not so unchangeable after all.

4.2 WHAT DO WE MEAN BY DATA REPRESENTATION?

We will then go on to look at a special way of representing large quantities of data with a certain type of structure, and the kind of 'arithmetic' which follows from this representation. Finally, in case the rest of the chapter has seemed a little abstract, we will see how the ideas we have been studying can be applied in a practical way.

4.2 WHAT DO WE MEAN BY DATA REPRESENTATION?

Before we go any further we need to sort out exactly what is meant by a representation of data. You may be wondering why all the fuss — how does the representation of the data differ from the data itself? And do different representations of the same data behave in different ways?

You can begin to see the distinction between the data itself and the way it is represented by thinking about the following situation. Anyone who has watched a cat with its four kittens knows that cats can count — at least, the cat will be uneasy if one of the kittens wanders out of sight, and recognizes when all four are near at hand. The 'data' which the cat is processing is the 'four-ness' of the kittens. No-one would suggest that she understands the concept of the word 'four', or the symbol 4; but at some level she recognizes the actual idea which 'four' or 4 represent.

At a slightly higher level, there is reason to believe that early in history people counted by matching sets with the same numbers of elements. We saw in Chapter 2 that the child's method of counting on fingers is a way of using a one-to-one matching between the set of fingers and the set of elements being counted; but archaeological evidence suggests that other methods were used when ten fingers were insufficient. Notches on sticks, for example, could be matched to cattle being counted, and it is not a great step from direct one-to-one matching to using some kind of shorthand for larger quantities — tens, for instance, or sevens or whatever. As we will see, this is more or less what happened. But the notches in the stick, like the child's fingers, are only representations for the abstract idea of 'four-ness'.

There are now many other, less simple ways of representing data, the most obvious one from your point of view being the use of patterns of magnetized material on the surface of a computer disc or tape — a long way removed from notches in sticks, yet only an extension of the same basic idea.

One further example emphasizes the difference between the data itself and the method we use to represent it. Consider temperature: the actual temperature — the underlying data — is the same whether we choose to represent it as 80°F or 27°C, though the representations look very different. We must take care, too, not to apply facts which are true of one system ('freezing takes place at zero') to another.

What, then, is the practical significance of this distinction between data and its representation? Everyone knows what is meant by 4, and as for the interpretation of the magnetic representation of numbers on a floppy disc, can't that be left to the software? We believe that there are in fact two reasons for giving a little more thought to the difference between the data itself and the way we choose to represent it. First, we must be careful not to take for granted as *features of the data* facts which are actually only the product of the way we have chosen to represent it, as the temperature example above shows.

Secondly, the ways in which familiar objects like numbers are represented may give us some clues when the time comes to devise representations of less familiar 'mathematical objects' such as the graphs and trees you will encounter later in the book. Often numbers in some form are the best way of representing non-numerical 'objects': a circle, for example, is not a 'number', but it can be specified by giving numerical values to certain features (its radius, the location of its centre with respect to some pair of (x,y) axes). However, these number representations will themselves vary depending on whether we measure in inches, centimetres, or some other unit.

4.3 WHAT KINDS OF NUMBERS ARE WE REPRESENTING?

Before we can discuss ways of representing numbers, we need to give some thought to the underlying objects being represented — the numbers themselves. We have already seen in Chapter 2 that numbers may be divided into several sets with different properties: the natural numbers $0, 1, 2, \ldots$; the integers $\ldots -3, -2, -1, 0, 1, 2, \ldots$; the rational numbers, which can be expressed as the ratio of two integers, such as 7/9 or 2/5; and the irrational numbers like $\sqrt{2}$ and $\sqrt{5}$, which cannot be so expressed. In Chapter 3 we actually proved that $\sqrt{2}$ is an irrational number, but there are many more. The decimal expansions of such numbers do not terminate (whereas the decimal expansion of 1/4, say, terminates at 0.25). Nor do they repeat themselves after a finite number of digits (as does the expansion of a number such as $2/7 = 0.285714285\ldots$). Instead, they have infinite, non-repeating decimal expansions — and thus, as we will see later in the chapter, cannot possibly be stored with total accuracy in any computer.

There exists also another class of numbers, e and π being examples, which are called *transcendental* numbers — they cannot be defined as the solution of any algebraic equation. These, too, have decimals which are infinitely long and non-repeating.

All these numbers together make up what is known as the *real number system*. It is helpful to think of the real numbers as strung out along a line, with zero in the middle, positives to the right, negatives to the left, and extending to infinity in both directions. The rational numbers are, as it were, extremely 'crowded' along this line — between any two rational numbers, however small the distance between them, we can always insert another (for example, between 5/6 and 7/8 we can put $(5+7)/6+8)$, and this method will always produce a number between two other rational numbers).

But there are also an infinite number of irrational numbers between any two rationals — a rather harder concept to visualize. The numbers, rational and irrational together, are spread in a continuous spectrum along the entire real number line from $-\infty$ to $+\infty$. It is this infinite set which we must find ways of representing.

4.4 THE DECIMAL NUMBER SYSTEM

We begin with the most familiar system of data representation — the decimal system. As you probably already realize, the system is called 'decimal' because it is based on powers of ten — a basis which arguably arose because we have ten fingers. It is also what is known as a *place value system*, which means that the position of a digit within a decimal number determines the value which is to be associated with that digit.

An example will make this clearer. For a decimal system, only ten digits are required (we will see why in a moment), and so we can make do with 0, 1, 2, ..., 9. When we write a number such as 324, this is really a shorthand for $10^0 \times 4 + 10^1 \times 2 + 10^2 \times 3$. In other words, the *place* or position of the digit, counted from the right, tells us by what power of ten it should be multiplied (always remembering that the first place represents 10^0, the 'units' term). Likewise if we have to deal with numbers less than 1, we simply introduce negative powers of 10 on the same basis; thus $0.041 = 10^0 \times 0 + 10^{-1} \times 0 + 10^{-2} \times 4 + 10^{-3} \times 1$. The decimal point marks the point in the string of digits between 10^0 and 10^{-1}.

You can now see why 10 digits are enough here; we can write 99 to mean $10^0 \times 9 + 10^1 \times 9$, but the next larger number, 100, uses the next place column, and returns to the start of our sequence of digits, so we never need a digit higher than 9. Later we will see that this can be generalized to tell us that in a place-value system based on powers of k, we need only k different symbols.

We are so accustomed to using the decimal system that we hardly ever explicitly think about its properties, and yet we use many of them implicitly in performing the operations of ordinary everyday arithmetic. For example, when we were educated (some time before you!) we learned, in carrying out our subtraction sums, to say 'borrow one and call it ten' when the digit to be subtracted was bigger than the one it was being subtracted from. We were (though we did not realize it at the time) 'borrowing' one of the tens from the next place-value position; we could only 'call it ten' because our system happened to be based on powers of ten.

You can probably think of other examples for yourself — the fact that to multiply by ten we just 'add a nought', or move the decimal point one place to the right, both of which operations have the effect of increasing all the place values by one power of ten; the fact that a number whose digits add to a total divisible by three must itself be divisible by three, and so on. (The proof of this last is quite entertaining: suppose we have a number *abc*, such as 426, whose digits add up to a multiple of three. That is, $a+b+c=3k$ for some integer k. The number *abc* is really a shorthand for $c+10b+100a=(c+b+a)+9b+99a=3k+9b+99c$, which divides by three.)

What we must be sure of, then, is that we do not try to take over with us into other methods of number representation features which are essentially related to this *particular* method. The features which *can* be taken over, as long as we remain within the real number system, are the properties which we summarized on page 35 in Chapter 2 — properties such as the commutativity of addition and multiplication, and the existence of a zero such that $a+0=a$. We can say that these are properties independent of the system of representation of the numbers — that is, properties inherent in the underlying numbers themselves.

4.5 NON-DECIMAL REPRESENTATIONS OF NUMBERS

If you have studied computing for any length of time you are probably already familiar with several other systems of number representation, in particular the binary system. This, like the decimal system, is a place-value system, but one which uses a different *base*, (this is the name given to the number whose powers determine the place values of the digits). In the decimal system this base was ten; in the binary system it is two, so that in line with what we said earlier, only two digits (usually 0 and 1) are required to write binary or base-two numbers. If we write 110 in this system (read as one-one-zero and not 'one hundred and ten' — we must get away from decimal terminology!) what we mean is $2^0 \times 0 + 2^1 \times 1 + 2^2 \times 1 =$ decimal $2+4=6$. If there is a danger that we may confuse numbers written using different bases, we sometimes clarify by writing $110_2 = 6_{10}$, but the meaning is often clear from the context.

Two other systems which are also met with in computing are the octal or base-eight system, and the hexadecimal or base 16 system. In this last, 16 symbols are required, and so as well as the familiar digit symbols 0, 1, 2, ..., the first six letters of the alphabet are used to correspond to decimal numbers 10, 11, ..., 15. Thus a hexadecimal number 4B would represent $16^0 \times 11 + 16^1 \times 4 =$ decimal 27. You can sometimes see a hexadecimal system in use in the way some floppy-disc formatting programs refer to disc sectors — you will see strings of characters such as 23A6F being printed on the screen while the formatting is going on.

Changing numbers back and forth between bases is a fairly sterile kind of activity, and not one which you would actually need to perform very often. We give just a few examples to convey the ideas.

Example 1
Express 370_{10} in binary.

Answer
Another way of putting this is that we want to split 370 into a sum of powers of two. What is the highest power of two which will divide into 370? (This section is best read with a calculator in hand!) A little experimentation will show that it is 256. We then continue to divide the remainder by the next lower power of two until we reach a remainder which is less than two, thus:

$$\begin{aligned}
370 &= 256 \times 1 + 114 \\
&= 256 \times 1 + 128 \times 0 + 114 \\
&= 256 \times 1 + 128 \times 0 + 64 \times 1 + 50 \\
&= 256 \times 1 + 128 \times 0 + 64 \times 1 + 32 \times 1 + 18 \\
&= 256 \times 1 + 128 \times 0 + 64 \times 1 + 32 \times 1 + 16 \times 1 + 8 \times 0 + 4 \times 0 + 2 \times 1 \\
&= 2^8 \times 1 + 2^7 \times 0 + 2^6 \times 1 + 2^5 \times 1 + 2^4 \times 1 + 2^3 \times 0 + 2^2 \times 0 + 2^1 \times 1 + 2^0 \times 0 \\
&= 101110010_2.
\end{aligned}$$

Of course, once you are accustomed to the process, you will not need to write it out in such detail. If you feel that the final number is very long, notice that any number greater than 2 will have at least three binary digits in its expression, numbers greater than 4 will have at least 4 digits, and in general numbers greater than 2^n will have at least $n+1$ digits. (The same is true, replacing 2 by 10, for decimal numbers, but we are so accustomed to this that we don't usually think about it.)

Example 2
Express 434_{10} as a hexadecimal number ('hex' for short).

Answer
The process for turning 434_{10} into hex is exactly the same as that for converting to binary, except that successive divisions are by powers of 16 instead of powers of 2. Thus:

$434 = 256 \times 1 + 178$

$$= 256 \times 1 + 16 \times 11 + 1 \times 2$$
$$= 16^2 \times 1 + 16^1 \times 11 + 16^0 \times 2$$
$$= 1B2_{16},$$

remembering what we said above about the use of A, B, ... F to represent 10, 11, ..., 15.

Example 3
Convert 716_8 to base 10.

Answer
This is easy — place-values in octal are powers of 8, so 716 means $8^0 \times 6 + 8^1 \times 1 + 8^2 \times 7 = 462$ in decimal.

Writing algorithms to carry out these conversions is more complex than might at first appear; we give an action diagram in Fig. 4.1 for converting binary to decimal.

```
┌─ Read B
│  Count number of digits in B=n
│  D=2^(n-1)
│ ┌ While n>1
│ │    Let n=n-1
│ │    Read nth digit of B, x_n
│ │    D=D+x_n × 2^(n-1)
│ └ End while
└─ Print D
```

D denotes the decimal version of the number, and B the binary version

Fig. 4.1.

Exercise 4.5.1
What is $4E11_{16}$ in decimal?

Exercise 4.5.2
Convert 67_8 to a binary number. Can you extend the process to give an algorithm for general octal-to-binary conversion?

4.6 SOME NON-PLACE-VALUE SYSTEMS

The Roman empire lasted for more than 1000 years, and produced great architects, generals, poets and civil engineers, but hardly any mathematicians or scientists. This contrasts with other ancient civilizations such as those of Greece, China, India and the Arab lands, all of which contributed substantially to the development of mathematics as we know it today. The difference could be largely due to the Romans' cumbersome way of representing numbers.

Roman numerals are not a true place-value system, though the position of a character within the number as a whole does have some significance. For example, in the Roman numeral MCMLXXXVIII, representing 1988, the C to the left of the second M indicates that the 100 (represented by C) is to be subtracted from the 1000 (represented by M). Had it been written on the right, it would have been added. The full number is made up as shown:

```
  M      C     M    L   X   X   I I I
1000 − 100 + 1000 + 50 + 10 + 10 + 10 + 1 + 1 + 1
```

With a system as complex as this, it is much more difficult to devise effective algorithms for performing arithmetic — even though the underlying quantities being represented are the same. Which only goes to show the importance of *effective* data representation.

Another, rather ingenious, system is used on the faces of some popular brands of watch. This is a system based on colours, so that 'one' might be represented by a blue square, 'two' by a red square and 'three' by a yellow one. Then, in order to prevent the introduction of too many colours, four=yellow and blue (=3+1), five=red and yellow, and so on. The trouble with this system, which includes no concept of multiplication and therefore allows us to represent numbers only as sums and not as sums of powers (as in a place-value system), is that the expressions for 4, 5, etc., are not unique. Four could just as well be expressed as red plus red, and even trying to apply some criterion such as 'fewest possible elements' is not helpful — both ways of showing 4 involve the same number of squares. Clearly this kind of ambiguity would spell disaster in any number system intended for the performance of useful calculations.

4.7 A FINITE NUMBER SYSTEM

All the ways of representing numbers that we have looked at so far have dealt with infinite sets of numbers, even though they managed to do so by using only a finite number of symbols. We look now at a system in which the actual set of numbers we need to deal with is finite.

Suppose we divide a number by, say 7. What are the possible remainders? The remainder cannot be bigger than 6, for if we had a remainder of 7, we could take out another factor of 7. So the set of remainders when dividing by seven is a finite set, consisting of 0, 1, 2, 3, 4, 5, 6 — seven possibilities in all.

We now agree to identify in some way all numbers having the same remainder when divided by 7; for instance, 45 and 66 would be identified according to this rule, since both have remainder 3 when divided by 7. We do not use an = sign for this identification, since this would risk confusion with the = of ordinary arithmetic. Instead, we write 45≡66 (mod 7). The ≡ sign is read as 'is *congruent* to' and the (mod 7) is an abbreviation for 'modulo 7', indicating that we are talking about remainders on dividing by 7. The whole statement tells us precisely that 45 and 66 have the same remainder when divided by 7. Similarly, 17≡5 (mod 4), and so on.

This *modulo arithmetic*, as it is called, is useful when we want to carry out a process such as selecting from a set of numbers those which are multiples of 4. We choose all those numbers which are congruent to 0 (mod 4) — that is, they have 0 remainder when divided by 4.

We can also carry out arithmetic which is entirely confined within the finite set of possible remainders, as follows. To add 6 and 4 (mod 7) we say 6+4=10, as usual, and then $10 \equiv 3 \pmod 7$, so $6+4 \equiv 3 \pmod 7$. Thus the results of all arithmetic operations can be expressed, in modulo n arithmetic, in terms of the finite set of remainders 0 to $n-1$. You are used to carrying out this kind of arithmetic modulo 12, probably quite unconsciously, whenever you make calculations involving times. With a 12-hour-clock system, we say that a three-hour exam starting at 10 will finish at 1 — in other words, $10+3=13 \equiv 1 \pmod{12}$.

We will not pursue the uses of this kind of arithmetic any further at present; it has very important applications in coding theory, which you will be reading about in Chapter 10.

Before we leave the topic of representation of numbers, we would like to emphasize that all the methods we have looked at are not just amusing distortions of the 'real' number system based on tens, but are valid systems in their own right, with their own useful applications (for example, the 'clock-face arithmetic' mentioned above). There is nothing special about the decimal system except the fact that it is the one most of us are accustomed to using. Had we been born at a different time, or in a different place, some other kind of system might have been equally second nature to us.

Exercise 4.7.1
Find $6 \times 7 \pmod{11}$.

Exercise 4.7.2
How could you express 'the set of all numbers of the form $6k+4$' in terms of modulo arithmetic?'

4.8 THE ACCURACY OF NUMBERS

A problem which often bothers students, particularly when studying subjects such as statistics, is 'How many figures should I quote in my answer? This problem is bad enough when you are using a calculator with an eight-figure display, but when calculations are being performed on a computer which can offer many more digits in its results, it becomes critical. It is therefore very important to realize that *no* computer carries out all its calculations with total accuracy, and that in some cases the results of in-built inaccuracy can have quite startling effects on the answers you obtain.

An amusing demonstration of the kind of inaccuracy we are talking about is provided by the factorial function on a scientific calculator. You probably know that the symbol 6! represents a *factorial*, and is shorthand for $6 \times 5 \times 4 \times 3 \times 2 \times 1$ (it's a concept we'll be using in Chapter 9). If you work out $6!/6-5!$ you should therefore get zero — and indeed you almost certainly will, if you use a calculator to perform the arithmetic. However, $n!$ becomes large very quickly as n gets bigger, and at some point it becomes difficult for the calculator to store the results of the computations accurately enough. The effect is so pronounced that, on the very reputable brand of calculator which we use, the result of calculating $65!/65-64!$ is -5×10^{79}! (the last ! is not a factorial).

What is going on inside the calculator — and inside a computer, for the processes are similar, though on a different scale — to produce this effect? The answer requires us to explore the way a computer represents numbers in a little more detail.

Many calculators and some computer procedures produce output in what is called *scientific*, or *exponent*, notation: a number such as 3479 would appear as 3.479E3, which is short for 3.479×10^3. The number is written in a form where there is only one digit before the decimal point, and is then multiplied by the appropriate power of ten (referred to as the *exponent*). In the same way 0.0065 would become 6.5E-3.

A slight variant on this is the *floating point* format, in which the number is written as a figure with the decimal point in front of the first digit, rather than after it (again multiplied by the necessary power of ten). In this format 3479 would become 0.3479×10^4. To store a number in this format requires several separate pieces of information to be stored — the actual number in its 0.XXXX form, the power of ten, plus its sign, and the overall sign of the number. Storing a large number can thus take up quite a lot of storage space; for example, to store -54973 would require us to express it as -0.54973×10^5, and the five digits 5, 4, 9, 7, 3, plus the exponent 5, and the overall minus sign, would all need to be stored, taking up seven storage locations altogether (in fact the number would probably be expressed in a different base, such as binary, but the principle is the same).

Given that storage space in any computer is limited, then, some loss of accuracy is bound to occur. Suppose that we needed to store the number -54973 in a computer which only had a maximum of four storage locations available for the digits of the number; then it would have to be stored as -0.5497×10^5, and would be inaccurate to the extent of three units, or about 0.005%. Not a very terrible inaccuracy, but such errors can accumulate as we carry out calculations involving many inaccurately stored numbers. If our computer could only store two digits, then 2.84 would have to be stored as 2.8, and if it is the square of this number which we want to find, the answer will be 7.84 instead of 8.0656 — an error of nearly 3%.

Moreover, if a number cannot be stored accurately, most computers do not 'round off' in the way you were taught to do. For example, if you want to reduce 2.89 to two significant digits, you would doubtless write it as 2.90, since that is the two-digit number closest to 2.89. Computers, however, tend to *truncate* numbers — that is, they simply chop off the excess digits, so that 2.89 would become 2.8, leading to a much larger error.

The devising of good numerical procedures, which give accurate answers to certain types of computation even when some inaccuracy of this kind is inevitable, constitutes an important area of mathematics known as numerical analysis. We will not be giving the topic any further attention here, but simply warn you to be wary of believing all the apparently significant figures which are produced at the end of a computerized calculation.

You need to be careful, too, about the order in which you carry out computations — for example, if you need to find $M \times N/P$, where M, N and P are all very large numbers, it may be safer to compute M/P first, obtaining a smaller number, and then multiply by N, rather than do the $M \times N$ multiplication first and risk getting what is called *overflow* — a number too big to be stored in the computer's memory (this is what happens if you try to divide by zero on your calculator). On the other hand, if M/P is going to be extremely small, it may cause *underflow* — a number too small for the computer's memory to deal with. Such a number will be stored as zero, and of course the result of the entire calculation will then also appear — erroneously — as zero.

Below are a few examples of this sort of situation which you can try using your pocket calculator; though the number of figures stored in a calculator is, generally speaking, far smaller than that which a computer would store, the same ideas will apply to both.

Exercise 4.8.1
Take the set of numbers 1, 2, 3, 4, 5. Find their sum, and call it T. Find also the sum of their squares, and call it S. Now compute the quantity $S/5-(T/5)^2$. The result should be $\sqrt{2}$ or 1.4142136. Do the same for the set 0.001, 0.002, 0.003, 0.004, 0.005. As these are equal to the first set of numbers divided by 1000, the answer should be 0.0014142136, or 0.00141421 if your calculator has an eight-digit display. What do you find?

Exercise 4.8.2
Calculate $743\times 52/6000$, rounding off to two significant figures at each stage of the calculation, and performing the calculation in the order (a) $(743\times 52)/6000$; (b) $743\times(52/6000)$; (c) $(743/6000)\times 52$. What do you notice? Which method is the most accurate and why?

4.9 THE REPRESENTATION OF SETS OF NUMBERS

Many of the problems for which computers are used require them to handle not merely single figures but whole sets of figures of the same type. A firm of stockbrokers may need to store on its computer lists of the numbers of shares which constitute its customers' portfolios, together with their current prices, and while this kind of data could be stored simply as a list, it is more efficient, as we will see, to store it in a form with rather more structure. This enables us not only to store and access the data, but also to manipulate it en masse, instead of having to perform many individual computations.

One such structured storage method, which you may already have met in your study of programming languages, is an *array*. The simplest form of array should remind you strongly of the *n*-tuples which we met in Chapter 2, since it is virtually the same thing. If we wish to store details of the number of shares in six different companies which make up Customer A's portfolio, we can do so by means of the 1×6 array $[n_1\ n_2\ n_3\ n_4\ n_5\ n_6]$. The '$1\times 6$' refers to the fact that the array has one row, six columns. It is sometimes abbreviated still further, to $[n_k]$, $k=1, 2, 3, 4, 5, 6$. As for the individual entries in the array, they can be denoted by n_k, or by $n[k]$ in which case k is a *subscript* or an *array parameter*. This notation enables us to refer to an individual item within the array without having to list the whole array.

Suppose now that we have a second 1×6 array which contains the prices of Customer A's shares. We will write this as $[p_1\ p_2\ p_3\ p_4\ p_5\ p_6]$. Then it makes sense to ask whether, in addition to storing the data in this way, we can carry out arithmetic with it to tell us, say, the total value of A's portfolio, or what will happen if all the prices are subject to a 10% increase.. When we carry out this type of arithmetic, we usually refer not to arrays but to *matrices* (singular *matrix*), and so before going any further into the details of such arithmetic, we need a few more definitions.

4.10 MATRIX DEFINITIONS AND TERMINOLOGY

A matrix is a two dimensional arrangement of numbers such as

$$\begin{bmatrix} 2 & -3 \\ 1 & 2 \end{bmatrix} \text{ or } \begin{bmatrix} 1 & 2 & 3 \\ 0 & 4 & -1 \end{bmatrix}.$$

We refer to a matrix as a 2×3 matrix if it has two rows and three columns (like the second example here). More generally, an $r \times c$ matrix has r rows and c columns (if you have difficulty remembering whether the first figure represents the number of rows or the number of columns, use the mnemonic — given to us by a student — 'rhubarb and custard'!). The numbers r and c are referred to as the *dimensions* of the matrix. The arrays we dealt with in the last section were matrices which happened to have only a single row.

An individual entry within a matrix is called an *element*, and the notation a_{ij} is often used to denote the element in the ith row and the jth column. Thus in the 2×3 matrix

$$\begin{bmatrix} 2 & 3 & 4 \\ 0 & 1 & -1 \end{bmatrix},$$

$a_{12} = 3$, $a_{21} = 0$, and so on. We tend to use capital letters for matrices, and to write $A = [a_{ij}]$.

We can define two matrices $A = [a_{ij}]$ and $B = [b_{ij}]$ to be equal if and only if $a_{ij} = b_{ij}$ for all values of i and j. An obvious consequence of this is that two matrices can only be equal if they have the same dimensions; to put it formally, if A is an $m \times n$ matrix, and B is a $p \times q$ matrix, and if $A = B$, then $m = p$ and $n = q$.

There are a number of special types of matrix which will be useful later. A *square* matrix is, as you would expect, a matrix with the same number of rows as columns — an $n \times n$ matrix, in other words. The line of elements running from top left to bottom right of such a matrix, as indicated here

$$\begin{bmatrix} 2 & 3 & 0 \\ 1 & 1 & 4 \\ 0 & 7 & 0 \end{bmatrix},$$

is called the *leading diagonal* of the matrix, and we refer to the *diagonal elements* 2, 1, 0. Of course, only square matrices can be said to have diagonals.

A matrix with elements *only* on the leading diagonal, such as

$$\begin{bmatrix} 1 & 0 \\ 0 & -1 \end{bmatrix},$$

is a *diagonal matrix*; one with a 'mirror-image' pattern of numbers reflected in the leading diagonal is a *symmetric matrix*, thus:

$$\begin{bmatrix} 6 & 0 & 1 \\ 0 & 4 & 2 \\ 1 & 2 & 7 \end{bmatrix}.$$

Finally, given any matrix $A = [a_{ij}]$, we can define its *transpose* A' as $[a_{ji}]$. This means that rows and columns are interchanged; for example, the matrix

$$A = \begin{bmatrix} 1 & 0 & 2 \\ 2 & 3 & 0 \end{bmatrix}$$

would have transpose

$$A' = \begin{bmatrix} 1 & 2 \\ 0 & 3 \\ 2 & 0 \end{bmatrix}$$

Some of this terminology will not be used until later chapters; meanwhile we return in the next section to the question of carrying out arithmetic with matrices.

Exercises 4.10.1
State whether each of the following matrices is (a) square, (b) diagonal, (c) symmetric

(i) $\begin{bmatrix} 1 & 3 & 5 \\ 0 & 2 & 0 \\ 0 & 0 & 7 \end{bmatrix}$ (ii) $\begin{bmatrix} 1 & 3 & 5 \\ 3 & 2 & 0 \\ 5 & 0 & 7 \end{bmatrix}$

(iii) $\begin{bmatrix} 1 & 2 \\ 2 & 1 \\ 0 & 1 \end{bmatrix}$ (iv) $\begin{bmatrix} 3 & 0 \\ 0 & 11 \end{bmatrix}$

Exercise 4.10.2

If $A = [a_{ij}] = \begin{bmatrix} 2 & 4 \\ 3 & 2 \\ 7 & 10 \end{bmatrix}$, what is

(i) a_{22} (ii) A'?

4.11 MATRIX ARITHMETIC: ADDITION AND SCALAR MULTIPLICATION

We start with the simplest operation — addition. Can we attach any meaning to the sum of two matrices? To see how this question might be answered, let's return to the problem of Customer A and his share portfolio. For simplicity, we will reduce the number of companies in the portfolio to 3, and say that he owns 50 shares in company 1, 60 in company 2 and 100 in company 3. This can be written $H = [50\ 60\ 100]$ (commas between elements are optional), where H denotes the share-holding matrix.

If he now buys a further 10 shares in company 1, 20 more in company 2, and does not change his holding in company 3, the matrix showing the change in his holding could be written $C = [10\ 20\ 0]$. We need to include the zero change here to make it quite clear that we are still talking about the same set of three companies.

Now if we want to show his new total holding, the obvious notation is to write new holding matrix $H_1 = [50\ 60\ 100] + [10\ 20\ 0]$, and to define this sum to be equal to $[50 + 10\ 60 + 20\ 100 + 0] = [60\ 80\ 100]$. Addition is thus defined on an element-by-element basis, with the corollary that only matrices of the same dimensions can be added. For example, we can add

$$\begin{bmatrix} 1 & 2 \\ 2 & 0 \end{bmatrix} + \begin{bmatrix} 0 & -1 \\ 1 & 3 \end{bmatrix}$$

to get

$$\begin{bmatrix} 1 & 1 \\ 3 & 3 \end{bmatrix},$$

An extension of this idea gives us a definition for the product of a matrix by an 'ordinary number'. If A doubles his holding of all three types of share, we could write his new holding as $[50\ 60\ 100] + [50\ 60\ 100] = [100\ 120\ 200]$, but by analogy with the operations of ordinary arithmetic it makes more sense to write this as $2 \times [50\ 60\ 100]$. In the context of matrix arithmetic we refer to 'ordinary numbers', such as the 2 in this operation, as *scalars* or *scalar quantities*, to distinguish them from matrices.

To sum up thus far: we have shown that $[a_{ij}] + [b_{ij}] = [a_{ij} + b_{ij}]$, and that $[ka_{ij}]$, where k is a scalar and a_{ij}, b_{ij}, denote the elements of matrices A, B, as explained above. You can verify for yourself that with these definitions addition is commutative and associative, (see p. 35), and that scalar multiplication is distributive over matrix addition.

4.12 MATRIX ARITHMETIC: MULTIPLYING MATRICES

Being able to multiply a matrix by a scalar is not really a complete answer to the question, 'Can we multiply matrices?' The real problem is to determine a definition of multiplication which will enable us to attach a useful meaning to $[a_{ij}] \times [b_{ij}]$, at least for some pairs of matrices.

To shed some light on how this might be done, we return to the share portfolio problem, and suppose that the shares whose holdings are given by the matrix [60 60 100] have prices given by [1.2 1.5 2.3], where the elements are given in pounds. We could try to define multiplication on an element-by-element basis, as we did above with addition, getting [60 90 230] as an answer — a sort of 'value matrix'. But a quantity of more interest is the overall value of the share portfolio, which ordinary arithmetic tells us is $50 \times 1.2 + 60 \times 1.5 + 100 \times 2.3 = 380$. Motivated by this fact, we try defining the product of a $1 \times n$ and an $n \times 1$ matrix in the same way:

$$[50\ 60\ 100] \begin{bmatrix} 1.2 \\ 1.5 \\ 2.3 \end{bmatrix} = [50 \times 1.2 + 60 \times 1.5 + 100 \times 2.3]$$

The fact that we have written the second — price — matrix as a column or $n \times 1$ matrix rather than as a row, may seem a little baffling at this stage, but it will become clear why we had to do this when we look at the way this method can be extended to larger matrices.

You can see that with this definition, the product of a 1×3 matrix and a 3×1 matrix is a matrix consisting of a single element — that is, a 1×1 matrix. This in turn suggests that there may be some restriction on the dimensions of matrices which can be multiplied. For example, with this definition we could not multiply [1 2] by

$$\begin{bmatrix} 1 \\ 0 \\ 3 \end{bmatrix};$$

there are not enough columns in the first matrix to 'match' with the rows of the second.

The question of the *compatibility*, as it's called, of matrices under multiplication becomes clearer when we try to extend our definition to larger matrices. Suppose that the stockbroker of our example has another client, with holdings in the same companies as A, but with different numbers of shares in each. Her holding might be shown as [40 70 10]. However, share price increases will affect both these clients in the same way, and so it is reasonable to ask whether we can carry out the computation of the total value of these two clients' portfolios in a single operation.

Our first step is to write the single matrix

$$\begin{bmatrix} 50 & 60 & 100 \\ 40 & 70 & 10 \end{bmatrix}$$

to show the holdings of both clients at once. In this 2×3 matrix, rows represent clients, columns the three companies. Now we want to multiply this 2×3 matrix by the 3×1 matrix

$$\begin{bmatrix} 1.2 \\ 1.5 \\ 2.3 \end{bmatrix}.$$

Clearly the answer should show the total value of the two clients' holdings, and it is sensible for rows in this product to represent the clients, just as in the first matrix.

$$\begin{bmatrix} 50 & 60 & 100 \\ 40 & 70 & 10 \end{bmatrix} \begin{bmatrix} 1.2 \\ 1.5 \\ 2.3 \end{bmatrix} = \begin{bmatrix} 380 \\ 176 \end{bmatrix}$$

is therefore a possible definition for the product of the two matrices — certainly it has a sensible practical interpretation.

This is in fact the definition of matrix multiplication which is generally used (though others, like the element-by-element version, have also been suggested). As we suspected, the definition carries restrictions as to the sizes of matrices which can be multiplied. Here we had, in terms of dimensions, $(2 \times 3) \times (3 \times 1) = 2 \times 1$, and the multiplication was only possible because the first matrix had three columns and the second had three rows. More generally, we can say that *any* two matrices are *compatible for multiplication* if the first has the same number of *columns* as the second has *rows*, and with this restriction can say that $(m \times n) \times (n \times p) = m \times p$, the letters simply representing the dimensions of the respective matrices. This is an extremely useful fact to remember, enabling us to write down the 'shape' of the product matrix before we have actually calculated any of its entries.

Now let's try to multiply two more matrices, neither of which consists of a single row or column. We will extend the share price problem to include valuation of the two clients' portfolios, not only at the current prices as represented by the matrix

$$\begin{bmatrix} 1.2 \\ 1.5 \\ 2.3 \end{bmatrix}$$

but also at last year's prices of

$$\begin{bmatrix} 1.1 \\ 1.3 \\ 2.3 \end{bmatrix}.$$

If we write a new 3×2 matrix to show both old and current prices, thus

$$\begin{bmatrix} 1.1 & 1.2 \\ 1.3 & 1.5 \\ 2.3 & 2.3 \end{bmatrix},$$

then in order to obtain the old and current valuations of the shares we want to perform the multiplication

$$\begin{bmatrix} 50 & 60 & 100 \\ 40 & 70 & 10 \end{bmatrix} \begin{bmatrix} 1.1 & 1.2 \\ 1.3 & 1.5 \\ 2.3 & 2.3 \end{bmatrix}.$$

Note that these two matrices are compatible for multiplication by our previous criterion, since a 2×3 matrix times a 3×2 matrix should give a 2×2 result. Moreover, the practical interpretation of the matrices also leads us to expect a 2×2 answer, since we can show the meaning of the rows and columns of the matrices in a formal way thus:

		Companies			Prices		
Clients		1	2	3	Old	New	Companies
	A	X	X	X	X	X	1
	B	X	X	X	X	X	2
					X	X	3

$$= \text{Clients} \quad \begin{matrix} \\ A \\ B \end{matrix} \begin{matrix} \text{Values} \\ \begin{matrix} \text{Old} & \text{New} \end{matrix} \\ \begin{bmatrix} X & X \\ X & X \end{bmatrix} \end{matrix}$$

We perform the arithmetic of the matrix multiplication according to exactly the same process as we used above, treating each column of the second matrix in the multiplication as if it were a separate 3×1 matrix. The first column of our 2×2 answer is therefore given by

$$\begin{bmatrix} 50 & 60 & 100 \\ 40 & 70 & 10 \end{bmatrix} \begin{bmatrix} 1.1 \\ 1.3 \\ 2.3 \end{bmatrix} = \begin{bmatrix} 50 \times 1.1 + 60 \times 1.3 + 100 \times 2.3 \\ 40 \times 1.1 + 70 \times 1.3 + 10 \times 2.3 \end{bmatrix} = \begin{bmatrix} 363 \\ 158 \end{bmatrix}$$

The second column is given by the multiplication already carried out above. Thus if we write the complete multiplication, it looks like this.

$$\begin{bmatrix} 50 & 60 & 100 \\ 40 & 70 & 10 \end{bmatrix} \begin{bmatrix} 1.1 & 1.2 \\ 1.3 & 1.5 \\ 2.3 & 2.3 \end{bmatrix} = \begin{bmatrix} 363 & 388 \\ 158 & 176 \end{bmatrix}.$$

The lines indicate the terms which are involved in arriving at the top left element of the product.

Writing down a general rule for matrix multiplication is messy and not particularly enlightening. The easiest way to remember the process is that the entry c_{ij} in the product $AB = C$ is obtained by multiplying the ith row of A by the jth column of B, in the way defined at the start of this section.

You need a good deal of practice in matrix multiplication to become really comfortable with the process, so here are two worked examples, followed by an exercise for you to try.

Example 1

$$\begin{bmatrix} 0 & 2 \\ 2 & 1 \end{bmatrix} \begin{bmatrix} 1 & -1 & 4 \\ 0 & 3 & 2 \end{bmatrix} = \begin{bmatrix} 0 & 6 & 4 \\ 2 & 1 & 10 \end{bmatrix}$$

Example 2

$$\begin{bmatrix} x & 2 & 0 \\ 1 & x & 1 \end{bmatrix} \begin{bmatrix} y \\ 2 \\ 3y \end{bmatrix} = \begin{bmatrix} xy + 4 \\ 4y + 2x \end{bmatrix}$$

— the algebraic quantities are handled in exactly the same way as the numbers.

Exercise 4.12.1
Find

$$\begin{bmatrix} 2 & 0 & 0 \\ 0 & 2 & 1 \\ 1 & 2 & 2 \end{bmatrix} \begin{bmatrix} 0 & 1 \\ 1 & 0 \\ 2 & 1 \end{bmatrix}$$

Exercise 4.12.2
Find

$$\begin{bmatrix} 2 & 0 & 0 \\ 0 & 2 & 0 \\ 0 & 0 & 2 \end{bmatrix} \begin{bmatrix} a & b \\ c & d \\ e & f \end{bmatrix}.$$

What is the effect of this matrix multiplication?

Exercise 4.12.3
Find

$$\begin{bmatrix} 0 & 0 \\ 0 & 0 \end{bmatrix} \begin{bmatrix} x \\ y \end{bmatrix}.$$

What can you conclude?

Exercise 4.12.4
What is

$$\begin{bmatrix} 2 & 3 \\ 1 & 4 \end{bmatrix} \begin{bmatrix} 0 & 2 \\ 2 & 1 \end{bmatrix} ?$$

What is

$$\begin{bmatrix} 0 & 2 \\ 2 & 1 \end{bmatrix} \begin{bmatrix} 2 & 3 \\ 1 & 4 \end{bmatrix} ?$$

What do you notice?

A number of facts should have come to light as you worked those exercises. The first is that a matrix of the form

$$\begin{bmatrix} a & 0 & 0 \\ 0 & a & 0 \\ 0 & 0 & a \end{bmatrix},$$

which has the same number everywhere on the diagonal, and zeros elsewhere, has the effect, when it multiplies another matrix, of simply multiplying every element of the second matrix by a constant factor a. This, you will recall, is identical with the effect of multiplying a matrix by a scalar. For this reason, matrices of this special form (and corresponding larger ones) are called *scalar matrices*.

Next, it is clear that a matrix of any dimensions, all of whose entries are zeros, will always give a product matrix consisting entirely of zeros, no matter what compatible matrix it is multiplied by. This is analogous to the role of zero in ordinary arithmetic; accordingly such a matrix is called a *zero matrix*.

Finally, and perhaps most strikingly, you will have discovered from Exercise 4.12.4 that, unlike multiplication of ordinary numbers, matrix multiplication is *non-commutative* — that is, given any two matrices A and B, in general $AB \neq BA$. A little thought should show that this clearly follows from the rules for matrix multiplication compatibility — a 2×3 and a 3×4 matrix can be multiplied (giving a 2×4 result), but if the order of multiplication were reversed, the product would be $(3 \times 4) \times (2 \times 3)$, which cannot be evaluated. In fact only pairs of matrices whose dimensions are of the form $p \times n$ and $n \times p$ can be multiplied in either order — and even then the answer will depend on the order of multiplication, being a $p \times p$ matrix in one case and an $n \times n$ matrix in the other. Only if the matrices are both square is there even a chance that they commute.

Here, then, is an example of the kind of situation about which we have been warning you — a case where the rules of 'ordinary arithmetic' cannot be transferred to the new arithmetic which we are constructing for matrices. When the matrices are written out in full it is perhaps unlikely that you would make such an erroneous assumption, but if you are working merely with letters, A, B, to represent the matrices, it is dangerously easy to do so.

By now you may be wondering when we are going to apply these new mathematical structures to the solution of some practical problems. The next two sections of this chapter accordingly concentrate on such applications.

4.13 MATRICES AND COMPUTER GRAPHICS

We saw in Chapter 2 that points in the Cartesian plane can be represented by ordered pairs $[x, y]$ of numbers, called the co-ordinates of the point. There is nothing to stop us regarding such an ordered pair as a 1×2 matrix, and operating with it as with any other matrix. For example, what happens if we take the point whose co-ordinates are $[1, 1]$, and multiply this little matrix by, say,

$$\begin{bmatrix} 3 & 0 \\ 0 & 3 \end{bmatrix} ?$$

The result is easily found to be $[3, 3]$ so that both co-ordinates of the original point have been multiplied by 3. This of course is to be expected, since

$$\begin{bmatrix} 3 & 0 \\ 0 & 3 \end{bmatrix}$$

is a scalar matrix. But it is when we re-interpret the result in graphical terms that a useful fact becomes apparent. If we take several points — such as the co-ordinates of the corners of the square $[0,0], [0,1], [1,1], [1,0]$ shown in Fig. 4.2, and multiply *each*

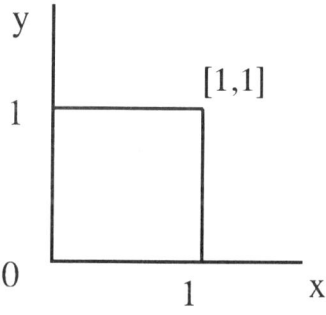

Fig. 4.2.

pair of co-ordinates by our matrix

$$\begin{bmatrix} 3 & 0 \\ 0 & 3 \end{bmatrix},$$

all the co-ordinates will be tripled simultaneously (except, of course, the zeros), and the square undergoes what we call a *magnification* — its linear dimensions are increased by a factor of three.

It is not hard to see that the same will happen with *any* scalar matrix with positive entries (though if the entries are less than one, the result will be a diminution rather than a magnification). Moreover, the idea could easily be extended to three dimensions — though that's not so easy to draw. Here, then, is a perfectly practical way of producing a magnification of a computer-graphic image: as long as the co-ordinates defining the image are known, a simple matrix multiplication will give the required magnification.

There is no reason why we have to restrict our investigation to scalar matrices with positive entries. What happens if we take the point [1,2] and multiply it by

$$\begin{bmatrix} 1 & 0 \\ 0 & -1 \end{bmatrix} ?$$

In this case we end up at $[1, -2]$, and the sketch in Fig. 4.3 shows that this is a reflection in the *x*-axis.

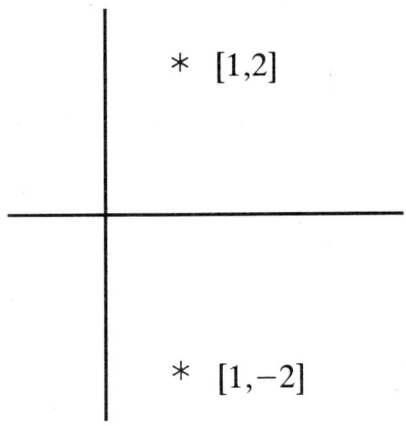

Fig. 4.3.

Exercise 4.13.1
What matrix multiplication would produce a reflection in the *y*-axis?

Exercise 4.13.2
What is the effect of multiplying the co-ordinates of points by

$$\begin{bmatrix} 2 & 0 \\ 0 & -2 \end{bmatrix} ?$$

As well as magnifications and reflections, images can be altered by rotating them — this is commonly required in computer-aided design when the engineer designing, say, a car-body panel wishes to look at it from several angles. If we take the point [1,0] and multiply by

$$\begin{bmatrix} 1 & 1 \\ 1 & 0 \end{bmatrix}$$

we get [0,1] — a rotation through 90° anticlockwise.

Exercise 4.13.3
What matrix multiplication would rotate images through 90° clockwise? (Hint: use the point [1,0] again, and ask yourself where it must end up when rotated.)

Of course we are approaching the problem of manipulating images from the opposite end to that which the designer of a graphics package would use. Rather than saying, 'What does this matrix multiplication do in geometric terms to my image?' he or she would say, 'What matrix multiplication would rotate my image by 30° clockwise?' To answer this question for angles other than 90° requires a knowledge of trigonometry which you may not possess, so we will not pursue it any further, but recommend anyone who is interested, and who has the necessary mathematical background, to refer to Scott's book in the 'Suggestions for further reading'.

The last type of geometric transformation we will examine is the effect, not of a matrix multiplication, but of addition. If we take the co-ordinates of our shape and add a constant 1×2 matrix, say [2,4], to each, the result will be to move the image 2 units to the right (parallel to the x-axis), and 4 units up (parallel to the y-axis). We call this a *translation* of the image; translations in other directions can be produced by varying the matrix $[a,b]$ which we add to the co-ordinates.

Naturally we are not restricted to carrying out these various geometric transformations one at a time. The matrix

$$\begin{bmatrix} 0 & 2 \\ 2 & 0 \end{bmatrix}$$

simultaneously magnifies by a factor of two and rotates through 90° anticlockwise. Thus complex transformations can be built up from the basic ones we have examined; some of these are investigated in the Problems at the end of the chapter. And once again, although the matrix arithmetic would be more complex, there is no reason why the same methods should not be applied to three-dimensional, or even multi-dimensional, images.

4.14 MATRICES FOR PROBLEM-SOLVING

Many applications of matrices, like some of the other techniques we have discussed, cannot be put into the form of a standard 'method'; you need to think through each problem individually. Here is an example to start you thinking along the right lines.

Pallets are used in large numbers by a transport company with depots in Coventry and Bristol. Examination of transaction records indicates that on average in any one week, 40% of pallets which begin the week in Bristol return there, while the remainder are transferred to Coventry. 50% of pallets starting out at Coventry return by the end of the week, while the other 50% are transferred to Bristol.

At the start of the first week in January 19XX, there are 500 pallets in Bristol, and 800 in Coventry. What will be the position at the end of the second week?

There is nothing in this problem which makes the use of matrix methods essential, but if we wanted to extend it to more realistic situations with many depots, calculations by the methods of ordinary arithmetic would become very cumbersome, and it would be more convenient, especially if dealing with the solution by computer, to use matrix methods. We will therefore express our simple problem in matrix terms.

However, some thought about the non-matrix way in which we would carry out the computation is useful. For example, to find how many pallets will finish the week in Bristol, we would need to calculate $0.4 \times 500 + 0.5 \times 800$ — and this looks very much like one element of a matrix product.

So we first write down a 2×2 matrix (call it T) to represent the transfer rates of pallets between depots, thus:

$$\begin{array}{c} \\ \text{From B} \\ \text{C} \end{array} \begin{array}{c} \text{To} \\ \begin{array}{cc} \text{B} & \text{C} \end{array} \\ \begin{bmatrix} 0.4 & 0.6 \\ 0.5 & 0.5 \end{bmatrix}, \end{array}$$

in an obvious notation. We now want to write another matrix N to denote the number of pallets starting the week at each depot, in such a way that either the multiplication NT or TN will give us the number at the depots at the end of the week. Clearly N must be either [500 800] or its transpose, and in the light of the calculation carried out earlier, we see that

$$[500 \ 800] \begin{bmatrix} 0.4 & 0.6 \\ 0.5 & 0.5 \end{bmatrix}$$
$$= [500 \times 0.4 + 800 \times 0.6 \quad 500 \times 0.6 + 800 \times 0.5]$$
$$= [600 \quad 700]$$

produces the correct results. (This is not the only way of arriving at the right solution; if instead of using NT we used $T'N'$, we would get the same answer, though in the form of a 2×1 instead of a 1×2 matrix.)

Since NT gives the numbers after one week, it follows that $NT \times T = NT^2$ gives those after two weeks — and indeed, the number after k weeks would be NT^k. After two weeks we will thus have

$$[600 \quad 700] \begin{bmatrix} 0.4 & 0.6 \\ 0.5 & 0.5 \end{bmatrix} = [590 \quad 710] \ .$$

Approaches of this kind are used in many applications — for example, in the study of how consumers 'brand-switch' from one manufacturer's product to another. You will find similar examples in the Problems at the end of the chapter.

4.15 MATRICES AND LOGIC

There is no reason why the elements of a matrix have to be drawn from the decimal number system. It is possible to have matrices whose elements are logical (Boolean) values T/F, on which we operate using the logical operators defined in Chapter 3. A simple example will serve to illustrate the idea.

If we have matrices

$$\begin{bmatrix} T & T \\ F & F \end{bmatrix} \text{ and } \begin{bmatrix} T & T \\ F & T \end{bmatrix}$$

whose elements consist of Boolean values, then we define the *matrix logical product* of the two by

$$\begin{bmatrix} T & T \\ F & F \end{bmatrix} \begin{bmatrix} T & T \\ F & T \end{bmatrix} \begin{bmatrix} (T \wedge T) \vee (T \wedge F) & (T \wedge T) \vee (T \wedge T) \\ (F \wedge T) \vee (F \wedge F) & (F \wedge T) \vee (F \wedge T) \end{bmatrix}$$

You can see that this process is completely analogous to ordinary matrix multiplication, except that we have ∧ (logical 'and') playing the role of multiplication, and ∨ (logical 'or') that of addition. By simplifying the elements of the product matrix in line with the rules established in Chapter 3, we find that it is equivalent to

$$\begin{bmatrix} T & T \\ F & F \end{bmatrix} \ .$$

You will be encountering applications of this idea to graph theory in Chapter 5, and a closely related set of matrices with binary elements will be used a great deal in Chapter 10 on coding theory.

We have only skimmed the surface of matrix algebra in this chapter; one important question which we have not covered is whether we can attach any meaning to 'matrix division' — in other words, if $AX = B$, where A, B, X are matrices, can we define a matrix A^{-1} — called the *inverse* if A — so that $X = A^{-1}B$, as we would say if we were solving an ordinary algebraic equation? The answer is a qualified yes, but if you wish to read more about the subject we suggest you get hold of the book by Griffel (see Suggestions for further reading).

PROBLEMS

1. Construct an action diagram or algorithm for the process of converting 24-hour clock times, such as 18.32, to 12-hour clock times. Test your formulation by inputting 5.27, 18.32 and 00.47.
2. If the computation $2597 \times 8362 \times 0.0493$ is carried out on a computational device which can only read, store and output numbers in the format $X \cdot X \times 10^n$, what will be the percentage error produced? What improvement in accuracy could be obtained by allowing $X \cdot XX \times 10^n$?
3. (a) A rev counter contains six resistors, four chips and four transistors. A strain gauge contains 5 resistors, six chips and one transistor. Set out this information as a 2×3 matrix E.
   (b) Resistors, chips and transistors can be bought from Germany or the USA. Costs are DM1.50, DM1.50 and DM2.00 respectively if the components are bought from Germany, $0.50, $0.10 and $1.20 if they are bought from the USA. Set out this information as two 3×1 matrices G and U.
   (c) If the current exchange rates are £1.00 = DM2.94 and £1.00 = $1.65, carry out scalar multiplications to convert G and U to new matrices G_1 and U_1 containing prices in pounds.
   (d) Carry out the two matrix multiplications EG_1 and EU_1 and interpret the results.
4. (a) Investigate the effect on the square whose corners are [0,0], [1,0], [1,1], [0,1] of multiplying each of these co-ordinates by the matrix

    $$\begin{bmatrix} 7 & 6 \\ 0 & 7 \end{bmatrix}.$$

    (Unless you know some trigonometry, this is most easily done by plotting the original and transformed squares accurately on graph paper.)
   (b) What is the effect on the cube with corners [0,0,0], [0,0,1], [0,1,1], [1,1,1], [0,1,0], [1,1,0], [1,0,1], [1,0,0] under multiplication of these co-ordinates by the matrix

    $$\begin{bmatrix} 0 & 2 & 0 \\ 2 & 0 & 0 \\ 0 & 0 & 1 \end{bmatrix}?$$

5. A chemical process involves three stages, mixing, heating and setting. One tonne of product A requires two hours of mixing, five hours of heating and six hours setting time, while a tonne of product B needs one hour mixing, 10 hours heating and four hours to set. Only one tonne of either product may be processed at a time.
   By matrix methods, determine the number of hours of each process required in a week when 10 tonnes of A and five tonnes of B are to be produced.
6. (a) How many multiplications are involved in carrying out the multiplication of a 2×3 and

a 3 × 4 matrix? (Multiplications are in general slower to carry out by computer than additions, so the addition operations can be neglected by comparison with the multiplications).
(b) Can you generalize what you deduced in (a) to the product of an $m \times n$ and $n \times p$ matrix?
(c) If you had to carry out the multiplication of a 3 × 8 matrix, an 8 × 5 matrix and a 5 × 2 matrix, which order of multiplication would be most efficient?

7. The $m \times n$ matrix $[a_{ij}]$ is to be stored in the form of an $mn \times 1$ array; for example, $\begin{bmatrix} 1 & 2 & 3 \\ 0 & 2 & 1 \end{bmatrix}$ will be stored as [1 2 3 0 2 1].
(a) Write an algorithm to read the matrix into the array.
(b) Write an algorithm to reconstruct the original matrix from the array, given the values of m and n.

5
Graph theory

5.1 GRAPHS — FOUR EXAMPLES

The word 'graph' may initially suggest some kind of line or curve drawn (usually) on squared paper — and this type of drawing is indeed correctly called a graph. Many such graphs represent mathematical equations such as $y=1.8x+32$: the drawing and the equation both represent the same relation between x and y. In this case the drawing corresponding to the equation would be a straight line.

You might notice that this particular equation has a further meaning beyond the mathematical meaning: if the quantity x represents a temperature measured in degrees Celsius, then the value of y would be the same temperature measured in degrees Fahrenheit. So we have an *algebraic* description (the equation) and a *pictorial* description (the graph), both of which model the connection between two scales of temperature measurement. We might think of the equation and the graph as alternative *mathematical models* of the relation between the two temperature scales.

Not all equations (and their related graphs) necessarily represent something else: we can also imagine equations and graphs which are purely abstract, and which don't represent anything outside themselves.

However, before we get too involved in this relation between mathematics and reality we will introduce our new sort of graph which in fact you already met briefly in Chapter 1. Firstly, it will have similar pictorial/mathematical properties to the familiar kind of graph mentioned above: we can draw it, and also represent it mathematically. Again, we will find that many of our graphs can be thought of as mathematical models of the relation between objects — using the word 'object' in a very wide sense. A graph, for our purposes, is a collection of points, called *vertices*, some of which are connected by lines, called *edges;* and as we are going to spend a lot of time considering the algebraic aspect of such graphs in later sections of the chapter, we will begin by looking at four examples of the pictorial representation. The first example is shown in Fig. 5.1 and concerns rail passengers.

One of the problems for long-distance rail travellers whose journey crosses

Sec. 5.1] Graphs — four examples

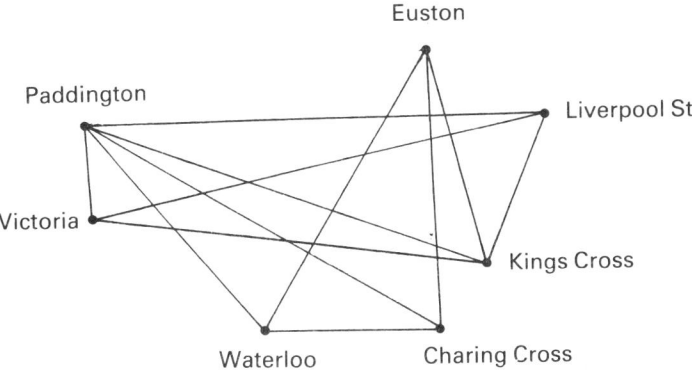

Fig. 5.1.

London is that all the main-line stations are termini, and it's necessary to travel across London by some means other than rail. Fortunately the London Underground connects all the stations, either directly or indirectly (by indirectly, we mean that it is necessary to change underground trains on a journey between two termini). We have produced a graph from the well-known map of the underground telling us which termini are *directly* connected. Vertices correspond to the termini, and if there is an edge between two vertices, it means that there is a direct connection between the corresponding stations. Notice that we have *not* drawn a map. The edges simply indicate the existence of a direct connection. On the original map Paddington, Charing Cross and Waterloo are stations on the same line — the Bakerloo Line as it happens. Our *graph*, on the other hand simply tells us that there is a direct connection from Paddington to Waterloo, without telling us that the route is through Charing Cross.

However, we include a word of warning: we should always bear in mind the limitations of any model we decide to use — after all, a model is usually the result of excising a lot of information. Consider how the following case might be misinterpreted. You will notice that there is no direct route from Charing Cross to Liverpool St.: the shortest route is certainly not the route via Paddington, which requires a single change. Our model includes only main line termini, which means that Tottenham Court Road station (which is where we *should* change trains) is not considered. In fact we aren't even modelling shortest routes: for all we know, there may be routes involving changes which are shorter and quicker than our direct routes! You can easily invent a map to demonstrate such a case.

Our second example is a very much simplified example of a food chain — something of interest to ecologists. Suppose we have the following information (some of which at least we hope you will agree with!). Rabbits eat grass; mice eat grain; foxes eat rabbits and mice; owls eat mice and beetles; beetles (eventually) eat rabbits, mice, owls and foxes. This information can be modelled by a graph, provided we put some directional information in — so that, for instance, grass doesn't eat rabbits. We will illustrate the *directional* nature of the relation 'rabbits eat grass and not the other way round' by modelling rabbits and grass as two distinct vertices, and then joining the two vertices by means of a *directed* edge or *arrow*. This is shown in Fig. 5.2: this type of graph is commonly called a directed graph, or a *digraph*.

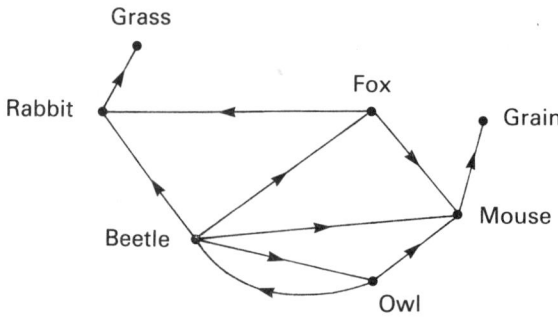

Fig. 5.2.

Although we are introducing the idea of a digraph here, we will not be giving it a full mathematical treatment until the next chapter: nevertheless it is worth your while to consider a couple of points at this stage.

For example, you might like to include an extra edge in the graph to the effect that mice are cannibals — mice eat mice. This is another possibility in any graph — a vertex is joined to itself.

Also, by reference to the digraph, you can trace the effect of the farmer using a poison to kill mice, which in turn poisons animals which prey on mice, and so on. You should be able to see that only the rabbits will be unaffected. On the other hand, if a poison fungicide were used on grass, you should be able to show that this would eventually affect all species except mice.

You have probably spotted that the direction of the arrows is the reverse of the direction in which effects are propagated through the food chain — and there is no reason why you should not model the reverse relation 'is eaten by', with the arrows reversed, if you think that on balance this is a clearer representation.

The next example is an example of a graph used in the social sciences to indicate the relationship between four factors — social factors, that is. Suppose an investigator has an idea that in any parliamentary polling district (a polling district is a section of a constituency) there is a relation between the following four factors:

(1) The percentage of the vote received by the government at the last election.
(2) The percentage of unemployed people in the polling district.
(3) The percentage of white collar workers in the district.
(4) The percentage of people classed as living below the poverty line.

These four factors can be represented by four vertices, with all four vertices

Sec. 5.1] **Graphs — four examples** 99

connected by edges, as shown in Fig. 5.3. This time the edges will be *labelled* with a plus or a minus sign according to whether the investigator thinks that there is a positive relation between the percentages (i.e. both tend to increase together), or a negative association (one tends to increase as the other decreases). So if the investigator thinks that there is a positive relation between unemployment and poverty rate, he or she will mark the corresponding edge with a plus sign. On the other hand, if it is thought that there is a negative relation between, say, unemployment and percentage of white collar workers — in simple terms, a high percentage of white collar workers in a constituency tends to go with a low percentage of unemployed — then the edge will be marked with a negative sign. Notice that we should strictly allow a neutral relation, where an edge is marked neither positive or negative: this case corresponds to a pair of factors between which the investigator thinks there is no relation, or doesn't know what the relation is likely to be. The edges on this graph don't have a *direction* because we haven't suggested that one factor is the *cause* of another factor.

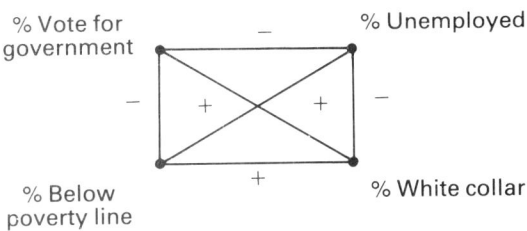

Fig. 5.3.

Even before the investigator examines the data from a sample of constituencies, he/she can test the *consistency* of the positive/negative markings. Why would a triangle of edges all marked with a negative sign be inconsistent? What about a triangle with two positive and one negative?

Now suppose that the investigator has data for, say, 20 widely differing polling districts. It is possible, by using statistical theory not covered in this book, to test the investigator's ideas quantitatively by calculating a measure known as the correlation coefficient, which measures the strength of relationship between a pair of factors. There will be six ways of selecting a pair of factors from the four under consideration (if you don't see why this should be, you will have to wait till Chapter 9 for a proof!). So six correlation coefficients can be calculated between the percentage occurrences of pairs of factors. The investigator's qualitative judgements about positive, negative and neutral relationships can then be compared with the correlation coefficients, which will themselves be positive, negative, or close to zero depending on the strength and nature of the relationships. In a serious study, the investigator could then proceed to test for causal relationships.

Our fourth model is intended to help us in the design of a road junction: a proposed new layout is given, together with the placing of traffic signals. The map in Fig. 5.4 shows a dual carriageway in an east–west direction, with a one-way road coming in from the south edge of the picture; and a two-way road from the north edge. The arrows represent the permitted directions from the signals: the signals are labelled with the letters *a* to *g*. Note that the map is itself a model of the junction.

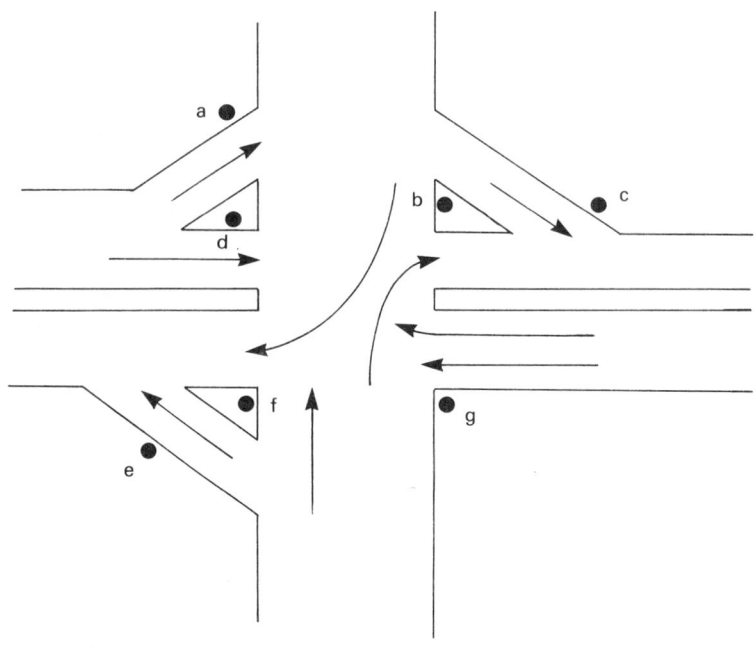

Fig. 5.4.

To model this map, we have a number of choices, but will look at just one. We will represent the traffic signals as vertices, using the same labels, a to g. If a pair of lights may both be green — allowing traffic to proceed *without* the traffic from one stream interfering with traffic from the other — then we will join the corresponding vertices with an edge. Look at lights a and b for an example. Where two streams *would* interfere — for example the streams from g and f — we don't connect the vertices, as this would correspond to putting two streams of traffic on collision course. Stated simply; if (and only if) an edge connects two vertices, then the corresponding lights may be green together. The graph is shown in Fig. 5.5 and we can see from the model that the possibilities for g, for example, are very restricted: the only other signal which may safely be green at the same time is signal c. What can you say about the set $\{a,d,e\}$? You might consider how to devise a sequence of signals with the help of the model — how can you maximise the traffic flow without putting any streams of traffic in danger? Imagine you have to allocate a two-minute cycle.

Finally you could experiment with an alternative graph. Suppose we connect a pair of vertices if it is *not* safe to allow the corresponding signals to be green at the same time, what does the new graph look like?

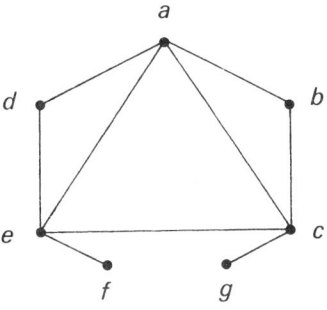

Fig. 5.5.

5.2 MATHEMATICAL REPRESENTATION OF GRAPHS

Now we need to look at the connection between the *graphical* representations we have just seen and the equivalent *mathematical* ideas. Remember how our straight-line graph (at the beginning of this chapter) had an associated equation: in a similar way we want to find mathematical models of our vertex-and-edge-type graphs. We will first show how to describe an undirected graph in terms of sets.

Example
The following set description gives rise to a graph as shown in Fig. 5.6

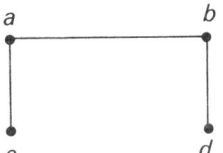 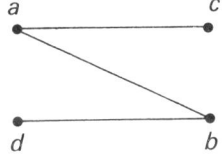 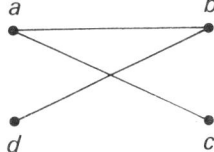

Fig. 5.6.

$G=[V, E]$ (graph)
$V=\{a, b, c, d\}$ (vertices)
$E=\{\{a,b\}, \{a,c\}, \{b,d\}\}$ (edge-pairs)

You should note that the three pictures represent the same information (in the same way that 128_{10} and 10000000_2 represent the same number) and we can choose whichever is most convenient for a given purpose.

We can think of a graph G in the following way:

$G=[V, E]$.

In words, the graph is an ordered pair consisting of V and E. Here

$$V=\{v_1, v_2,\ldots,v_n\},$$

i.e. V is a set with n elements, each of which represents a vertex.

E is a set composed of a number of two-element subsets of V, each subset representing an edge. For example, we might have

$$E=\{\{v_1, v_3\}, \{v_1, v_6\}, \{v_2, v_n\},\ldots,\{v_5, v_7\},\ldots\}$$

from which we could deduce that among other edges there is one connecting v_1 and v_3.

For convenience we will refer to the two-element subsets of V which constitute the elements of E as *edge-pairs*. Note that as these edge-pairs are themselves sets, they are not ordered.

Example
We are going to use a graph to model the relation 'is acquainted with' among five people. The information we have is that George is acquainted with Ira and Jane. Helen is acquainted with Jane and Ken. Ira is acquainted with Ken. We assume that acquaintanceship is a two-way affair, so the graph is undirected! The set description is

$G=[V, E]$
$V=\{$George, Helen, Ira, Jane, Ken$\}$
$E=\{\{$George, Ira$\}, \{$George, Jane$\}, \{$Helen, Jane$\},$
$\{$Helen, Ken$\}, \{$Ira, Ken$\}\}$

and this gives us the graph shown in Fig. 5.7. In passing you should note that we can choose to

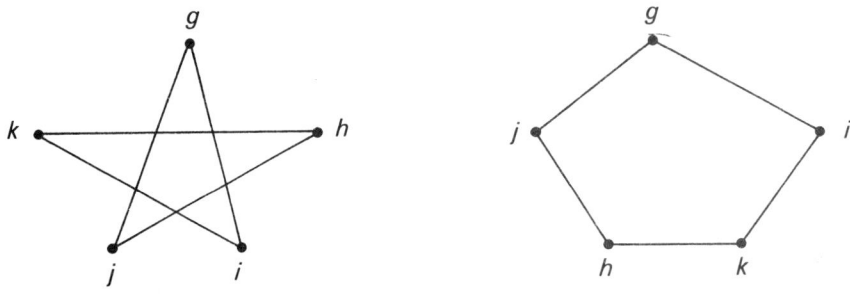

Fig. 5.7.

make the graph either star-shaped or pentagonal, as the shape makes no difference to the information contained in it. Also notice that we can adopt any clear labelling convention — initials for names in this case — as long as it does not contradict any other labelling conventions.

We will make frequent comparisons between the set notation for an undirected graph, and the corresponding pictorial representation of the same information. In other words we will examine the relation between certain characteristics of the ordered pair $G(=[V,E])$ and the properties of the corresponding graph. We have already seen two such examples: here are two more simple cases where features of the set representation correspond to quite specific pictorial (or graphical) effects.

Example 1
Suppose we take the representation

$G=[V,E]$
$V=\{a,b,c\}$
$E=\emptyset$.

Then pictorially we have three disconnected vertices a, b and c.

Example 2
On the other hand if we have

$G=[V,E]$
$V=\{a,b,c\}$
$E=\{\{a,b\},\{c,d\}\}$

we have an inconsistent description, as d is not in the set V; so we can't draw the graph.

Vertex degree
The next idea is a simple one, and allows us to take note of what we might call an 'honesty condition' or a necessary property of the set description of a valid graph.
Suppose we have a graph described as follows

$G=[V,E]$
$V=\{a,b,c,d,e,f,g,h\}$
$E=\{\{a,b\},\{a,d\},\{b,c\},\{b,d\},\{b,e\},\{d,e\},\{d,f\},\{e,g\},\{f,g\}\}$,

giving rise to the pictorial representation shown in Fig. 5.8.

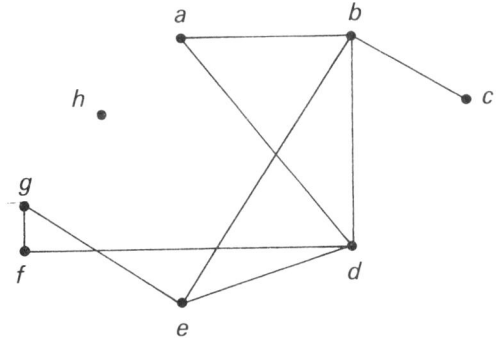

Fig. 5.8.

Graph theory

The *degree* of a vertex is equal to the number of edges meeting at it: so from the picture we can see that for example

deg(*a*)=the degree of vertex *a*=2
deg(*b*)=4
deg(*c*)=1
deg(*h*)=0.

In terms of the set description, we can think of the degree of a given vertex v, as being equal to the number of occurrences of the element v in the set of edge-pairs E: each occurrence means that an edge terminates at the vertex v.

a occurs twice
b occurs four times
c occurs once
h occurs not at all

Notice the following results: the sum of the vertex degrees in a graph must be an even number. This result is self-evident, if we note that the edge set E consists of pairs of vertices; and that each pair contributes exactly 2 to the total for the whole graph. (It is possible to obtain the same conclusion from consideration of a pictorial version of the graph: you may like to try it.) As stated above, an even sum of vertex degrees is a *necessary* property of a graph: if it doesn't have the property, then it isn't a graph. Because of this property we can note that, in any graph, the number of vertices of odd degree must be an even number. The result is of interest mainly because it allows us to demonstrate a very simple example of a proof by contradiction, such as we met in Chapter 3.

The four steps are shown below

(1) Assumption
   Assume that there exists a graph with an odd number of vertices of odd degree.
(2) Consequence of assumption
   It follows that the sum of vertex degrees for the graph is also odd, (because the sum of an odd number of odd numbers is odd).
(3) Fact
   But we have just shown that the sum of vertex degrees in any graph must be even.
(4) Consquence not consistent with Fact.

This contradicts our initial assumption: we conclude that there does not exist a graph with an odd number of vertices of odd degree. You should note that we are not able to give a drawing to demonstrate this proof. Any proof by contradiction which contains a diagram must necessarily be an optical illusion!

Sec. 5.3]   **The idea of isomorphism**   105

Exercise 5.2.1
Check the following two set descriptions for validity: test your conclusion by drawing!

(a) $G=[V,E]$
        $V=\{u,v,w,x\}$
        $E=\{\{u,v\},\{u,x\},\{v,x\},\{v,w\},\{w,x\}\}$

(b) $G=[V,E]$
        $V=\{a,b,c,d,e,f,g\}$
        $E=\{\{a,c\},\{b,c\},\{c,d\},\{a,d\},\{e,f\},\{g,d\}\}$

5.3 THE IDEA OF ISOMORPHISM

If you compare the graphs which you have drawn as solutions to some of the exercises in this chapter with those drawn by your colleagues, you may well find that the 'pictures' look quite different. Yet they may all correctly represent the same edge set/vertex set description of the graph. We say that the different pictorial representations are *isomorphic*, a word derived from the Greek iso=same, and morph=shape.

This term may at first sight seem odd, since the graphs are patently *not* the same shape in the ordinary physical sense; what the term refers to is the essential equivalence of the underlying structures being represented. The Oxford English Dictionary definition of isomorphism is 'an exact correspondence as regards the number of constituent elements and the relations between them'. We will take this definition as our starting point, and explore the idea a little with particular reference to graphs (you will come across other applications later in the book).

We say that two graphs, A and B, are isomorphic if we can put the vertex set of A into one-to-one correspondence with the vertex set of B, and can also put the edges defined by those vertices into a matching one-to-one correspondence. We can put this more formally as follows:

Two graphs, $A=[V_A,E_A]$ and $B=[V_B,E_B]$ are isomorphic if we can put the elements of V_A and V_B in one-to-one correspondence in such a way that $[x,y]\in E_A$ if and only if the corresponding vertex pair $[u,v]\in E_B$.

Note that this is a definition rather than an algorithm; it tells us what we must establish to demonstrate the existence of an isomorphism, but gives no instructions as to how the required facts can be established. (Contrast this with the definition of the connectedness property of graphs to be found in Section 5.4, which gives an algorithmic test for the property.) In fact it is usually much easier to show that two graphs are not isomorphic than to prove that they are. With graphs of any size, a great deal of sorting and comparison may be needed to establish the positive result.

To illustrate the definition, we will demonstrate that the two graphs illustrated in Fig. 5.9 satisfy the requirements given above, and thus are isomorphic. The set description of these graphs is

$G=[V_A,E_A]$, $G=[V_B, E_B]$
$V_A=\{a,b,c,d\}$, $V_B=\{e,f,g,h\}$
$E_A=\{\{a,b\},\{a,c\},\{a,d\},\{b,d\},\{c,d\}\}$,
$E_B=\{\{e,f\},\{e,h\},\{f,g\},\{f,h\},\{g,h\}\}$.

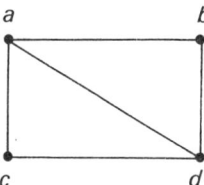
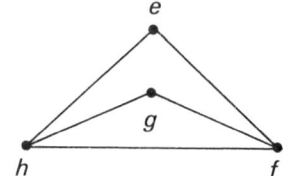

Fig. 5.9.

The necessary correspondences can be made as shown in Fig. 5.10 (though a certain amount of trial and error may be required to arrive at this conclusion).

We note in passing some necessary (and fairly obvious) conditions which must apply if two graphs are to be isomorphic:

(a) the two vertex sets must contain the same number of elements;
(b) the two edge sets must contain the same number of elements;
(c) the degree sequence of each graph must be the same (that is, if we list the vertex degrees for the vertices of each graph in descending order, we should get the same list in each case).

If any of the above conditions is not met, the two graphs *cannot* be isomorphic. However, since these are necessary and not sufficient conditions, we should be cautious about making the statement that two graphs are isomorphic even when these three conditions are satisfied. All three graphs in Fig. 5.11 have six vertices and seven edges. In all three cases there are two vertices which have degree 3 and four vertices which have degree 2. But no graph is isomorphic with any other.

Exercise 5.3.1
One of the following pairs of graphs is isomorphic; the other pair is not. As an exercise, identify which case is which, remembering that it is easier to show that two graphs do not constitute an isomorphism than that they do. In the positive case, find the correspondences of vertex elements and the correspondences of vertex-pairs (or edges), in order to show the isomorphism. Note that you will find this a lot easier if you use drawings of the graphs (shown in Fig. 5.12).

(a) $V_A=\{a,b,c,d,e\}$
        $E_A=\{\{a,b\},\{a,d\},\{a,e\},\{b,c\},\{b,d\},\{b,e\},\{c,d\},\{d,e\}\}$
   and
        $V_B=\{v,w,x,y,z\}$
        $E_B=\{\{v,w\},\{v,z\},\{w,x\},\{w,y\},\{w,z\},\{x,y\},\{x,z\},\{y,z\}\}$

(b) $V_A=\{a,b,c,d,e\}$
        $E_A=\{\{a,b\},\{b,c\},\{c,e\},\{c,d\},\{d,e\}\}$
   and
        $V_B=\{v,w,x,y,z\}$
        $E_B=\{\{v,w\},\{w,x\},\{w,z\},\{x,y\},\{y,z\}\}$

5.4 CONNECTEDNESS

This is a notion which is important in any system involving communications networks. For example, if the names of the five acquaintances in Section 5.2

Sec. 5.4] Connectedness 107

$V_a = \{a, b, c, d\}$

$V_b = \{e, f, g, h\}$

$E_a = \{\{a,b\}, \{a,c\}, \{a,d\}, \{b,d\}, \{c,d\}\}$

$E_b = \{\{e,f\}, \{e,h\}, \{f,g\}, \{f,h\}, \{g,h\}\}$

Fig. 5.10.

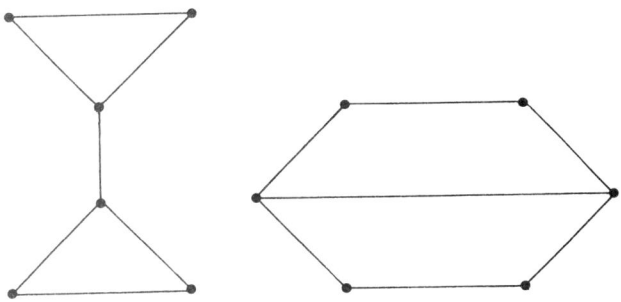

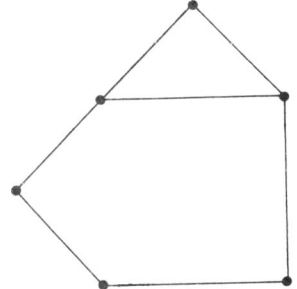

Fig. 5.11.

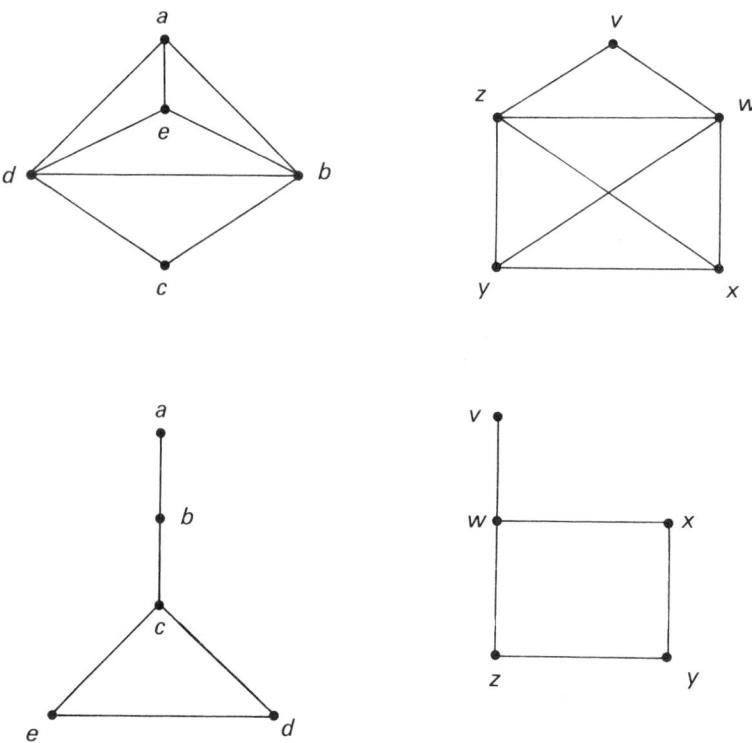

Fig. 5.12.

(diagrammed in Fig. 5.7) had been the names of five computers, and the relation had been that of connection between the computers, the system described would be a network of computers all of which are able to communicate with each other. In some cases the communication would be direct; in other cases indirect, via intermediate members of the network, but no computer would be isolated from all the others. (We'll see more examples of networks in Chapter 6).

With a relatively small vertex set V, it is possible to make a quick visual check for connectedness — we see a vertex with no edges associated with it; or we see that the graph is broken into two parts — then we can say that the graph is disconnected. What we require, however, is something more formal than this. We need a test procedure (i.e. an algorithm) to verify or refute the statement that a graph is connected: this procedure should use the given vertex and edge (V, E) information and should output a decision on that basis; after all, the graph will be stored in some such form.

Suppose we take our acquaintance graph (Fig. 5.7), with the vertex set

$V = \{g, h, i, j, k\}$

and edge set

$$E=\{\{g,i\},\{g,j\},\{h,j\},\{h,k\},\{i,k\}\}.$$

How can we determine from this specification that the graph is connected? We will carry out the process as follows:

(1) Choose an arbitrary vertex, which we define to be the first-level vertex.
(2) Find all vertices with a direct connection to the first-level vertex and list these as second-level vertices.
(3) Find all vertices, not already listed, with a direct connection to second-level vertices and list these as third-level vertices.
(4) Repeat process (3) for connections from third to fourth level, fourth to fifth level, and so on until no more new vertices can be found by this method.
(5) If the number of vertices found is equal to the number of elements listed in the original vertex set V, then the graph is connected.

Before we apply this to a specific example it is worthwhile recalling the meaning of the set representation. The set E is a set of edge-pairs. Each edge-pair is a two-element set, containing two distinct elements of the vertex set V. Each two-element set is a subset of the vertex set V. In these terms we can illustrate the process applied to our acquaintance graph (Fig. 5.7) as follows:

(1) Take an arbitrary element of V — say h — as the first-level vertex.
(2) Search the edge set E for edge-pairs which contain h. We find $\{h,j\}$ and $\{h,k\}$; so vertices h, j and k form a *connected subgraph* (we deal with subgraphs more formally in Section 5.8; just think of a subgraph as a piece of the original graph for the present!). We can now erase vertex h from set V and erase the edge-pairs $\{h,j\}$ and $\{h,k\}$ from E. Our second-level vertices are j and k.
(3) We next look at second level vertices j and k in turn. To find any third-level vertices connected directly with j, we look at the edge-pair set E and see that there is now just one element containing j, namely $\{g,j\}$, so vertex g forms part of a connected subgraph with j, k and h. We delete j from set V, and delete $\{g,j\}$ from set E, as before. The other second-level vertex was k. The edge-set element $\{i,k\}$ means that i is a third-level vertex and forms part of a connected subgraph with g, j, k and h. We delete k from set V and $\{i,k\}$ from E, and list third-level vertices g and i.
(4) We now consider the third-level vertices g and i. We see that E no longer contains any edge-pairs with either g or i as an element. (In fact at this stage E doesn't contain any elements at all.)
(5) This means that we have obtained the largest possible connected subgraph containing vertex h. Since the number of vertices in this subgraph, namely 5, is the same as the number of vertices in the original graph we can say that the original graph is connected.

5.5 PATHS

At this point it is worth taking note of the idea of a *path* in a graph. Pictorially it is just what the name suggests: any route from one vertex to another, which utilizes edges and associated vertices, is a path. We will restrict our discussion of paths to simple paths. A simple path is a sequence of edges which contains a given vertex at most once. The *length* of a path is defined as the number of edges in it. If we refer again to Fig. 5.7 and consider, say, the vertices g and h, we see that there are just two paths between the two vertices: the first consists of the edges $\{g,j\}$ and $\{j,h\}$, the other of $\{g,i\}$, $\{i,k\}$, and $\{k,h\}$.

A convenient way of representing such a path is by using a list, which as you know from Chapter 2 has the property of order, unlike a set; and also avoids the fixed-length characteristic of an *n*-tuple. The list contains the appropriate vertices in the same sequence as that in which they occur in the path. We can represent the path (of length 2) between h and g as

$P=\langle h,j,g \rangle$

A special kind of path is known as a *circuit*. A circuit is a path whose first and last vertex are the same.

Example
Look back at Fig. 5.12. Verify that the longest simple path in each graph is of length 4. Show that the longest *circuits* are of length 5, 5, 3 and 4 respectively.

5.6 BIPARTITE GRAPHS

Suppose we have a microcomputer with the following restrictions on its operating system. There are five processes, certain pairs of which may not take place at the same time. The five processes are described by the verbs (1) calculate (2) input (3) output (4) print and (5) store. The following processes may not take place in parallel: (1) calculate and output; (2) calculate and input; (3) output and store; (4) output and print.

We can set this out as the following graph.

$G=[V,E]$
$V=\{c,i,o,p,s\}$ (processes)
$E=\{\{c,i\},\{c,o\},\{o,p\},\{o,s\}\}$ (processes which may not run in parallel)

This formulation is shown in Fig. 5.13 in two different forms. Notice how much

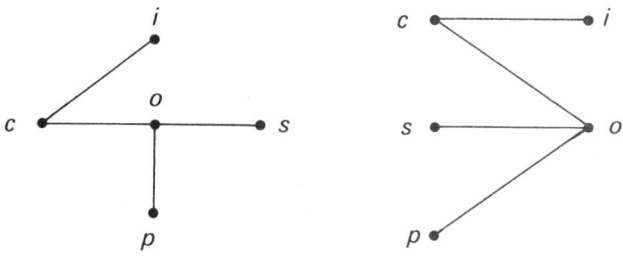

Fig. 5.13.

more informative the second version is: in this second drawing we see that the vertices fall into two subsets; and we can partition the vertex set V into two disjoint parts: $V_1=\{p,c,s\}$ and $V_2=\{o,i\}$. All edge-pairs in E contain one element from V_1 and one element from V_2. This property defines a *bipartite* graph; in the case of our imaginary computer the partition indicates that we need at most two separate time-slots to allow all of the processes to be carried out.

This is a very modest conclusion; nevertheless, the general idea of bipartition is a valuable one, and bipartite graphs are common, as we will see in subsequent work. (Another example of the notion arose in connection with isomorphism in Section 5.3, where the 'mapping' graphs (Fig. 5.10) were — amongst other things — bipartite.) Let us therefore give a formal definition:

A graph $G=[V,E]$ is defined as *bipartite* if it is possible to partition V into two disjoint subsets V_1 and V_2, in such a way that all edge-pairs in E contain one element of V_1 and one element of V_2.

From the definition we will see that it is not difficult to produce an algorithm to test a graph for bipartition. What makes it relatively easy is the 'either–or' nature of the classification of vertices: we should be able to *put* each vertex of the graph into V_1 or V_2 in such a way that the edge-pairs have the property described above. If this can't be done, the graph isn't bipartite.

The process has similar characteristics to the one used in Section 5.4 to establish connectedness. We will try it out informally with a small graph, before giving a more formal algorithm. The graph in Fig. 5.14 is clearly bipartite: how do we test it?

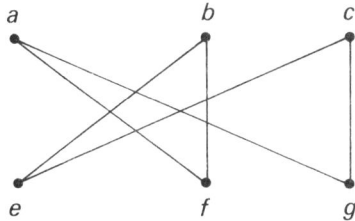

Fig. 5.14.

(1) Suppose we have two sets V_1 and V_2.
(2) We will arbitrarily place label a in set V_1.
(3) This means that we put f and g into set V_2, since f and g are adjacent to a. We note further that we have taken account of all vertices adjacent to vertex a, so we need no longer consider vertex a.
(4) Since f and g are in set V_2, b and c must go into set V_1 (and we need not consider f and g again).

(5) Since b and c are in set V_1, e must go into V_2.
(6) We have allocated all vertices to either V_1 or V_2, and our method of checking has accounted for all the edges of the graph. We can therefore state that the graph is bipartite: by our method of vertex examination, all edges have been checked. (Note that a single edge could be sufficient to invalidate the bipartite property: it could, for example, give rise to a triangle of vertices.)

You may notice that the process of establishing that a graph is bipartite corresponds to demonstrating the following in respect of its vertex set V:

$$V = V_1 \cup V_2 \text{ where } V_1 \cap V_2 = \emptyset.$$

In the present example $V_1 = \{a, b, c\}$ and $V_2 = \{e, f, g\}$.

We can determine whether or not a connected graph is bipartite by means of the following algorithm, which attempts to partition the vertex set of the graph into two sections. The instructions place each vertex of the graph in a section, and at the same time increase the value of count by one. If, at the end of the process, count exceeds n then at least one vertex has been placed in both sections, and the graph is not bipartite.

```
Entry
G[V,E] = a connected graph
V = {v₁, v₂, v₃, . . . vₙ} a set of vertices of G
E = {{vᵢ, vⱼ}: vᵢ, vⱼ V and vᵢvⱼ is an edge of G}
    S₂ = ∅
    S₁ = {v₁} (first vertex is placed in S₁)
    count = 1
    DO
        FOR i=1 to n
            IF vᵢ ∈ S₁ and vᵢ ∈ V
                FOR j=1 to n
                    IF {vᵢ, vⱼ} ∈ E and vⱼ ∈ V
                        S₂ = S₂ {vⱼ} (a vertex is placed in S₂)
                        E = E − {vᵢ, vⱼ} (an edge is removed from G)
                        count = count+1
                    ENDIF
                ENDFOR
                V = V − {vᵢ} (all edges from vᵢ have been checked)
            ENDIF
        ENDFOR
        exchange S₁ with S₂
    UNTIL E = ∅
    IF (count = n)
        graph is bipartite
    ELSE
        graph is not bipartite
    ENDIF
Exit
```

Sec. 5.6] **Bipartite graphs** 113

The process starts with the graph, $G=[V,E]$. Once an edge $\{v_i, v_j\}$ has been examined, its edge pair representation may be deleted from the set E, and once *all* the edges incident on a given vertex v_i have been examined, the corresponding element v_i in V may be deleted. Also — for simplicity — at the end of each pass through the outer loop, the two blocks of the partition have their labels interchanged, so that we are always looking from S_1 to S_2, when checking adjacencies. To show how this algorithm works, we will try it out on two small graphs G_1 and G_2 shown in Fig. 5.15, which differ by just one edge. We will write out changes which occur during each pass through the outer loop (the 'DO' loop). You should check the intermediate changes in the 'FOR' loops and satisfy yourself that the summarized results are correct.

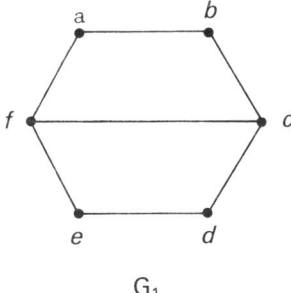 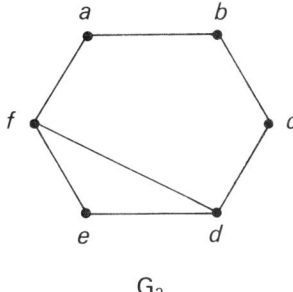

Fig. 5.15.

Example
Are the graphs G_1 and G_2 bipartite?

Answer

$$G_1=[V,E]$$

Initially

$V=\{a,b,c,d,e,f\}$
$E=\{\{a,b\},\{a,f\},\{b,c\},\{c,d\},\{c,f\},\{d,e\},\{e,f\}\}$
$S_1=\{a\}$ $S_2=\emptyset$ count=1

First call of 'DO'

$S_2=\{b,f\}$ $V=\{b,c,d,e,f\}$
$E=\{\{b,c\},\{c,d\},\{c,f\},\{d,e\},\{d,f\},\{e,f\}\}$
$S_1=\{b,f\}$ $S_2=\{a\}$ count=3

Second call of 'DO'

$S_2=\{a,c,e\}$ $V=\{c,d,e\}$
$E=\{\{c,d\},\{d,e\}\}$
$S_1=\{a,c,e\}$ $S_2=\{b,f\}$ count=5

Third call of 'DO'

$S_2=\{b,f,d\}$ $V=\{d\}$

$$E=\emptyset$$
$$S_1=\{b,f,d\} \quad S_2=\{a,c,e\}$$

Exit from loop (since $E=\emptyset$).
 The vertex set has been partitioned, since $S_1 \cap S_2=\emptyset$, and we see that the graph is bipartite.

$$G_2=[V,E]$$

Is this graph bipartite? We will apply the same algorithm
 Initially

$$V=\{a,b,c,d,e,f\}$$
$$E=\{\{a,b\},\{a,f\},\{b,c\},\{c,d\},\{d,e\},\{d,f\},\{e,f\}\}$$
$$S_1=\{a\} \quad S_2=\emptyset \quad \text{count}=1$$

First call of 'DO'

$$S_2=\{b,f\} \quad V=\{b,c,d,e,f\}$$
$$E=\{\{b,c\},\{c,d\},\{d,e\},\{d,f\},\{e,f\}\}$$
$$S_1=\{b,f\} \quad S_2=\{a\} \quad \text{count}=3$$

Second call of 'DO'
$$S_2=\{a,c,d,e\} \quad V=\{c,d,e\}$$
$$E=\{\{c,d\},\{d,e\}\}$$
$$S_1=\{a,c,d,e\} \quad S_2=\{b,f\} \quad \text{count}=5$$

Third call of 'DO'

$$S_2=\{b,f,d,e\} \quad V=\emptyset$$
$$E=\emptyset$$
$$S_1=\{b,f,d,e\} \quad S_2=\{a,c,d,e\} \quad \text{count}=8$$

Exit from loop (since $E=\emptyset$).
 The set V has not been partitioned, since $S_1 \cap S_2 \neq \emptyset$, so the graph G_2 is not bipartite.

5.7 VERTEX COLOURING

A bipartite graph is said to be 'two-colourable' since we need only two labels or 'colours' to mark vertices in such a way that no two connected vertices share the same colour. All the vertices corresponding to one part of the partition can be labelled as, say, 'red' vertices: the others as 'green' vertices. Such a labelling of the picture has exactly the same effect as the partitioning of the graph vertices V into the disjoint sets V_1 and V_2 but is so simple that anyone with two coloured pencils could carry out the exercise. If the graph is not bipartite, then at some point more than two colours will be needed.

The algorithm for bipartition, which was given in Section 5.6, tested whether a given graph (in set notation) was bipartite, and, if so, also gave the partitioning of the vertices. As you can see, the coloured pencil exercise and the more formal algorithm are equivalent. But now we want to look at cases of graphs which are *not* bipartite, so to keep our arguments simple, we will stick to the coloured-pencil-and-diagram approach.

Example
Verify that the two graphs in Fig. 5.16 are not two-colourable and hence not bipartite.

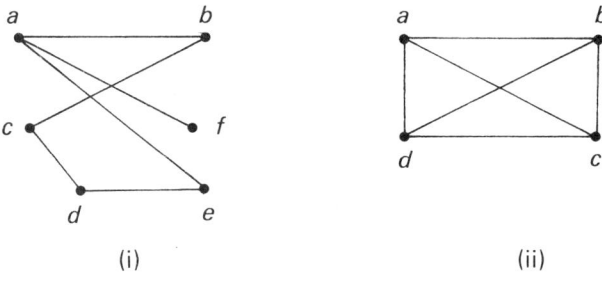

Fig. 5.16.

Although it may not be immediately obvious, the graphs drawn in Fig. 5.16 contain circuits in the form of a pentagon and a triangle respectively. Re-draw the graphs if you're not certain. Since neither the triangle nor the pentagon can be two-coloured, it is clear that the graphs containing the triangle and the pentagon cannot be two-coloured either. You should verify that (i) is three-colourable: obviously graph (ii) requires four colours, as all vertices are interconnected. Graph (ii) is an example of what is called a *complete* graph, which you have already met in Chapter 3 — in this case a complete graph on four vertices (often denoted by K_4). A partitioning (or colouring) for (i) is $\{a,d\}$ $\{c,e,f\}$ and $\{b\}$, while for (ii) it is simply $\{a\}$ $\{b\}$ $\{c\}$ $\{d\}$.

It can be shown that a *planar* graph requires at most four colours (a planar graph is a graph which *can* be drawn on a surface in such a way that none of the edges of the graph cross each other). This fact is closely related to the minimum number of colours needed to illustrate areas on a map — a problem which was mentioned briefly in Chapter 3. Let's look at this question in a little more detail.

Suppose we have an island containing six countries (and surrounded by sea!) shown in Fig. 5.17. How can we colour the map so that no two areas (countries or sea) which share a border line are given the same colouring?

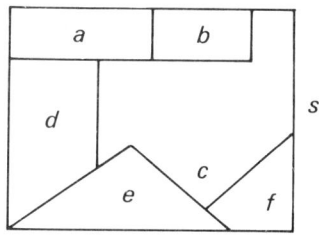

Fig. 5.17.

First we produce a graph model by representing each area on the map by a vertex on the graph. Second, if two areas share a border we draw an edge on the graph connecting the corresponding vertices. The resulting graph is shown in Fig. 5.18. (Notice how much information from the real situation we are discarding. But this is the purpose of a model — to highlight the features which concern us.) One possible colouring for the graph vertices is $\{s\}$ $\{c\}$ $\{a,e\}$ $\{b,d,f\}$: then the allocation of colours to *vertices* on the graph is the colouring for the corresponding *areas* of the map. Given the lack of a precise algorithm to carry out the colouring of vertices the best we can do is to make 'reasonable' choices, allowing that it may be necessary to backtrack if our colour scheme needs more than four colours, since theory tells us that four are sufficient. One possible first allocation of a colour is to the vertex with the highest degree, as this tends to simplify subsequent decisions. In the present case this is the vertex corresponding to the sea(s) or else the vertex c.

The idea of representing incompatibility of adjacent objects by graph has applications outside of geography: we modelled a simple microcomputer operating system in the previous section by such a graph, for instance; and you will meet other cases later.

Exercise 5.7.1
Look back at Fig. 5.5, which was a representation of *compatible* sets of traffic signals. Redraw the graph to represent *incompatible* sets of signals. Show that your graph is four-colourable, and that this will enable you to allocate a sequence of signals (as hinted at in Section 5.1!)

5.8 COMPLETE GRAPHS, SUBGRAPHS AND SUBDIVISIONS

In the last section you saw an example of a *complete graph* K_4. The general complete graph, K_n is defined as a graph with n vertices, each of which is joined by an edge to every other vertex. In terms of sets a labelled K_4 graph can be specified as

$G=[V,E]$
$V=\{a,b,c,d\}$
$E=\{\{a,b\},\{a,c\},\{a,d\},\{b,c\},\{b,d\},\{c,d\}\}$

The five smallest complete graphs are shown in Fig. 5.19. (Notice that K_4 is the largest complete graph which is also planar (see 'Subdivision' below).) We can also extend the idea to bipartite graphs, and define a *complete bipartite graph* $K_{m,n}$ as a bipartite graph which has the further property that all m vertices in one section of the partition are connected by an edge to all n vertices in the other section. So the complete bipartite graph $K_{3,2}$ for example, has three vertices in one section, and two in the other. Examples are shown in Fig. 5.20. You might like to verify that $K_{3,2}$ is planar, and that $K_{3,3}$ is not. (Is $K_{4,2}$ planar?) We will not produce an algorithm to test a set representation of a graph for completeness, though some further light will be shed on the question in Chapter 9. The main requirement for completeness in general is that the graph has the right number of edges; in the bipartite case, that it has the right number of edges and is also bipartite! The problem of calculating this 'right number of edges' for a given number of vertices, will be addressed in Chapter 9.

Sec. 5.8]   **Complete graphs, subgraphs and subdivisions**   117

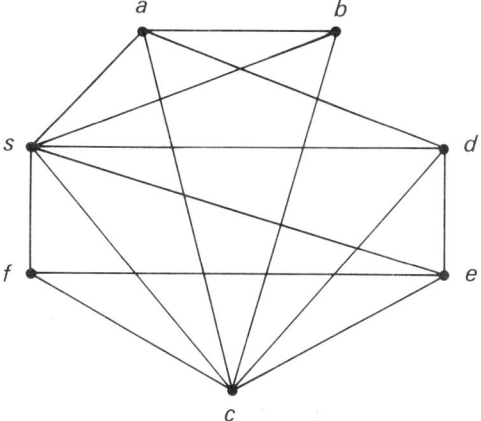

Fig. 5.18.

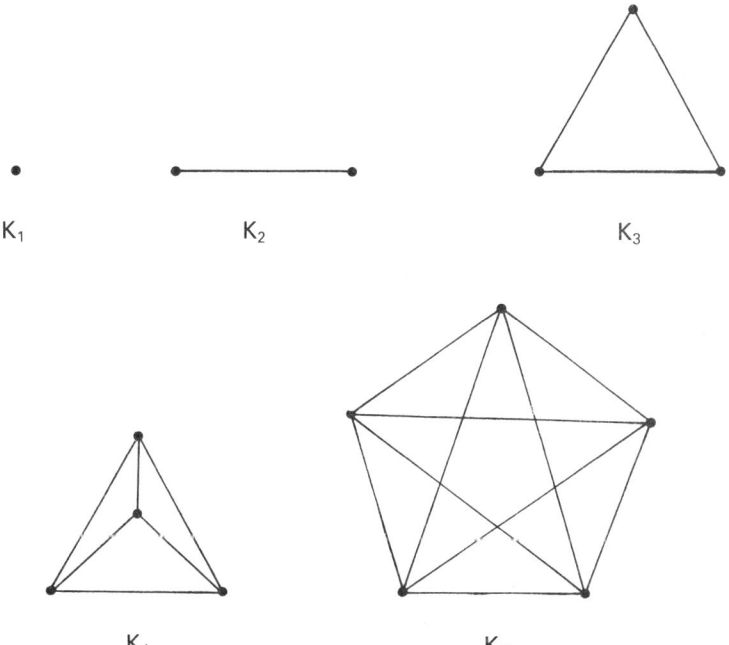

Fig. 5.19.

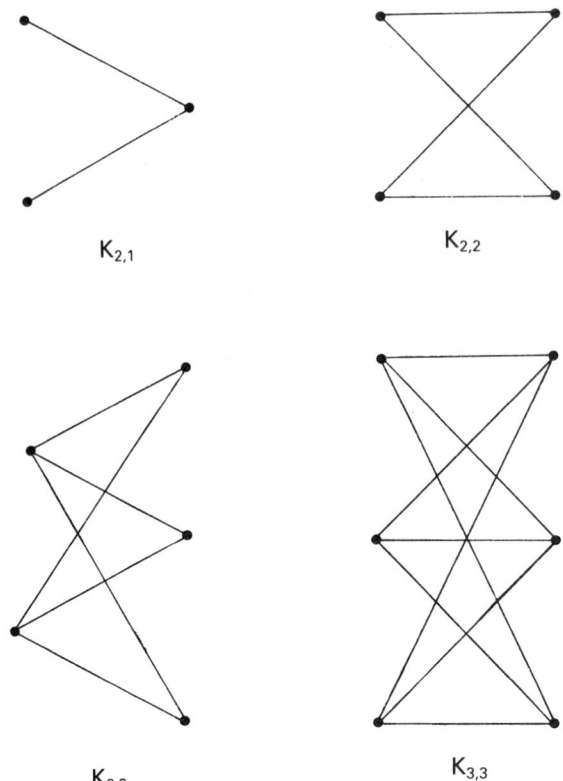

Fig. 5.20.

Subgraphs

The idea of a *subgraph* (which we met originally in Section 5.4) can be shown by drawing or by set description, though drawing will usually be sufficient for our purposes.

Basically we can say that

$$G_s = [V_s, E_s]$$

is a subgraph of

$$G = [V, E]$$

if $V_s \subseteq V$ and $E_s \subseteq E$

For example the graph shown in Fig. 5.21 can be represented in set terms as

$$G = [V, E]$$
$$V = \{a, b, c, d, e, f\}$$

Sec. 5.8]   **Complete graphs, subgraphs and subdivisions**   119

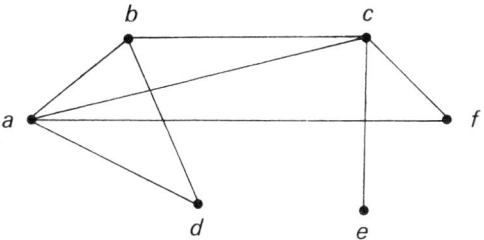

Fig. 5.21.

$E = \{\{a,b\},\{a,c\},\{a,f\},\{a,d\},\{b,d\},\{b,c\},\{c,e\},\{c,f\}\}$

An example of a subgraph of G is shown in Fig. 5.22, and has the set representation

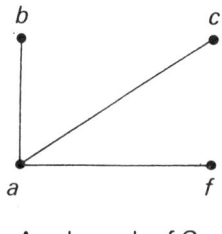   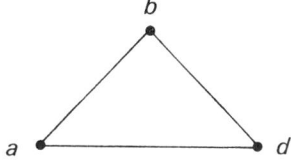

A subgraph of G   A complete subgraph of G

Fig. 5.22.

$G_s = [V_s, E_s]$
$V_s = \{a,b,c,f\}$
$E_s = \{\{a,b\},\{a,c\},\{c,f\}\}$.

Note that the elements of edge-pairs in E_s must consist of elements of V_s: edges in a graph exist only by virtue of vertices. There are several hundred possible subgraphs of G in this example: for a start there are 63 different ways of choosing a sub-set of V containing one or more vertices, before even considering the allowable edges. As it turns out, we need not concern ourselves with the majority of subgraphs: the important subgraphs are those which are *complete* subgraphs — and usually K_2 or larger.

As an example $V_s = \{a,b,d\}$

$E_s = \{\{a,b\},\{a,d\},\{b,d\}\}$

represents a complete subgraph G_s of G.

120   **Graph theory**   [Ch. 5

Exercise 5.8.1
(a) Find by inspection a K_4 subgraph in the graph G, which is drawn in Fig. 5.23.

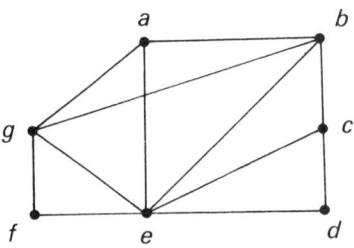

Fig. 5.23.

(b) Find a set of complete subgraphs which between them contain all the edges of G. (A vertex may be in more than one subgraph; as may an edge.)

Subdivision
Finally in this introduction to undirected graphs we will take a brief look at the idea of a *subdivision* of a graph. The main value of this idea for us is to help us to test whether a graph is planar or not: that is, whether it can be drawn in two dimensions with no edges crossing. For this reason we will treat the idea in terms of diagrams. A subdivision of a graph is produced when a graph is modified, in the following way. Take an edge of the graph. Replace the edge by a path of length 2. This replacement path includes a new vertex of degree 2. (In simple terms, this process is equivalent to placing a new vertex in the middle of an edge.) Note that any subdivision of a planar graph will also be planar: any subdivision of a non-planar graph will also be non-planar. (Again, in simple terms, sticking additional vertices onto edges won't change the planarity, or lack of planarity of the graph). We will leave this section by stating a theorem on planarity due to Kuratowski:

A graph is planar if, and only if, it does not contain a subgraph which is K_5 or $K_{3,3}$ (or a subdivision of K_5 or $K_{3,3}$).

Note that this theorem gives a necessary and sufficient condition for a graph to be planar. The *necessary* condition can be stated as: if a graph is planar, then it does not contain a subgraph which is K_5 or $K_{3,3}$ (or a subdivision of K_5 or $K_{3,3}$). Don't forget: a necessary condition for an event is a condition which always (of necessity) occurs in conjunction with the event. The 'event' in this case is 'the graph is planar'. (If you need to refresh your memory on necessary and sufficient conditions, refer to Chapter 3.) It also follows from this necessary condition, and from your knowledge of logical implication, that a graph containing a subgraph which is K_5 or $K_{3,3}$ (or a subdivision of these) is *not* planar.

If this necessary condition were a full statement of the theorem, it would not amount to much more than a statement of the obvious; after all we know that K_5 and

Sec. 5.8]   **Complete graphs, subgraphs and subdivisions**   121

$K_{3,3}$ are not planar, so any graph containing either of them as a subgraph will not be planar either. However the stronger part of the theorem is the 'sufficient' part. You should recognize by logical implication that the theorem tells us that *all* non-planar graphs contain a K_5 or a $K_{3,3}$ (or a subdivision of K_5 or $K_{3,3}$) as a subgraph. In other words, whenever we encounter a non-planar graph, we can find at least one such subgraph in it.

Exercise 5.8.2
Recall the graph G shown in Fig. 5.23. Find a plausible reason for there not being a subgraph which is either a K_5 or a $K_{3,3}$ (or a subdivision). Re-draw the graph to show clearly that it is in fact planar.

The subject of planarity is a difficult one for the non-specialist, though we might consider a simple example to illustrate a possible application. Suppose we have a set of five transistors, all directly connected. Obviously this can be represented by the complete graph K_5: vertices can represent transistors, and edges can describe the connections.

If we want to set this circuit up on a printed circuit board or lay it out on a microchip, we will need more than one layer, otherwise connections will cross. What we can do is to find a minimum set of connected subgraphs such that all subgraphs are planar. If we allocate one layer to each subgraph, it will then be necessary to connect (in the electrical sense) vertices which occur in more than one layer — the process is illustrated in Fig. 5.24.

As we know, K_4 is planar: so we can lift off a K_4 subgraph using the vertices (say) a,b,c, and d as part of the first subgraph. We can attach edges $\{b,e\}$, $\{a,e\}$, and $\{e,d\}$ to the K_4 graph and still have a planar subgraph. This means that the first layer may contain all the edges except $\{c,e\}$; and the single edge $\{c,e\}$ can constitute the second layer.

Exercise 5.8.3
In the above example, find an alternative allocation of edges to layers for K_5 so that one subgraph has five edges and the other has four.

At the time of writing there exist transputer chips containing up to 500 000 components, admittedly not all directly connected, so you will see that this subject is a very complex one. You might prefer to finish Section 5.8 in a slightly flippant way, whilst at the same time getting hands-on experience of planarity.

The game of sprouts is played between two players as follows. All that is needed is pencil and paper. Start by marking a small number of dots (vertices!) on the paper: three vertices is plenty for a beginner. Players take it in turn to join a pair of vertices by drawing an edge: and a new vertex is drawn somewhere near the middle of the new edge.

There are two restrictions to a player's choice: a new edge may not cross any existing edge; also the *maximum* number of edges which may meet at a vertex is three. (You can put these restrictions into technical terms, given your new knowledge of vertex degree and planarity.) The winner is the last player who can make a

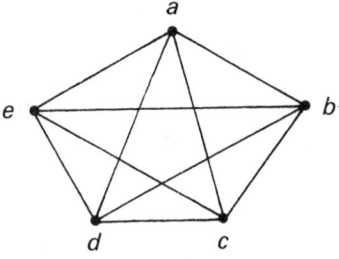

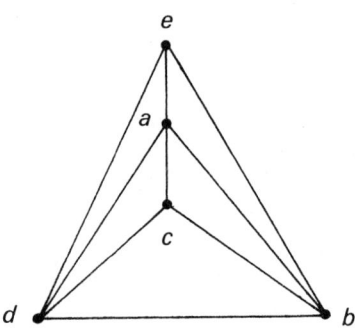

Fig. 5.24.

legal move. The history of the shortest possible game of sprouts — played on just *two* vertices — is illustrated in Fig. 5.25. You might like to try your own version on three or four vertices.

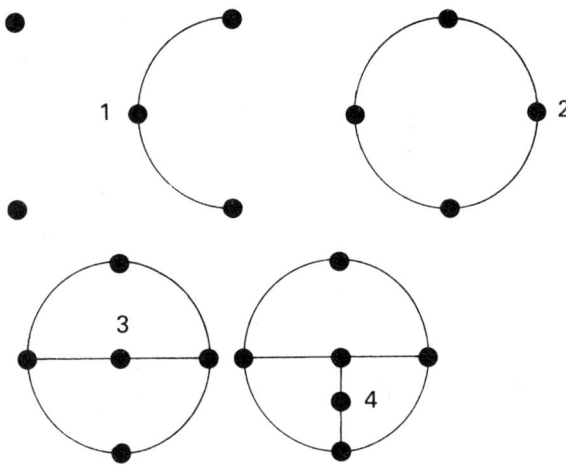

Fig. 5.25.

Verify that the first player in the two vertex game always loses!

Problems
1. Draw the graphs corresponding to the set descriptions below.
   (a) $G_1 = [V_1, E_1]$
       $V_1 = \{a,b,c,d,e,f\}$
       $E_1 = \{\{a,b\},\{a,f\},\{b,c\},\{c,d\},\{c,f\},\{d,e\},\{e,f\}\}$
   (b) $G_2 = [V_2, E_2]$
       $V_2 = \{u,v,w,x,y,z\}$
       $E_2 = \{\{u,x\},\{u,z\},\{v,x\},\{v,y\},\{v,z\},\{w,y\},\{w,z\}\}$
   (c) $G_3 = [V_3, E_3]$
       $V_3 = \{a,b,c,d,e,f,g\}$
       $E_3 = \{\{a,d\},\{a,c\},\{b,f\},\{b,e\},\{c,d\},\{d,g\}\}$
2. Write out the vertex-degree list for each graph given in Question 1.
3. Apply the algorithm given in Section 5.5 to the above set descriptions to show that G_1 is connected whereas G_3 is not.
4. Use the algorithm in Section 5.6 to show that G_2 is bipartite.
5. Identify (in list notation) the longest and the shortest circuits in G_1.
6. Establish whether or not G_1 is isomorphic with G_2: either give a reason why not, or show an isomorphism.
7. A small firm of systems analysts put all of its six software experts into an open plan office. Unfortunately it was found that certain individuals were incompatible with others, and it was decided to introduce smaller work rooms. The incompatible pairs were Andy and Barry, Andy and Frank, Barry and Clare, Dave and Eddy etc., as indicated by initial in graph G_1, where an edge indicates incompatibility.
   (a) Use the idea of graph vertex colouring to enable the firm to allocate congenial groups of software experts into the minimum number of rooms!
   (b) Gerry is to be added to the group. Unfortunately he is not compatible with Andy or Barry. Will another room be necessary in order to maintain harmony? Give a reason.
   (c) Draw up a set descripton of a new graph G_4 where vertices still correspond to individuals, but now the graph takes account of Gerry, and unlike the original incompatibility graph, an edge in G_4 indicates *compatibility* between pairs of individuals.
   (d) Draw a pictorial representation of the compatibility graph G_4. Choose a set of complete subgraphs from G_4, so that the firm can allocate compatible people to the minimum number of work rooms, and every individual shares with at least one other.

6
Digraphs

6.1 DIRECTED GRAPHS

So far in our applications of graph theory we have been looking at relations which are symmetric; we could just as truthfully have said that

(1) George is acquainted with Jane; or
(2) Jane is acquainted with George; or simply that
(3) George and Jane are acquainted:

and this was why we used a subset (which is not ordered) as the data type, when describing edges. Although the two-element subset was written as $\{g,j\}$, since we usually put names in alphabetic order, the subset $\{j,g\}$ is identical. Now we are going to take a look at non-symmetric relations; for example, George precedes Jane in the alphabet, or George is older than Ira; or Ken owes money to George, or Helen is a parent of Jane, and so on. In such cases, as Ken knows only too well, the order of the two elements in the relation is relevant, and we allow for this by modifying the set description which was introduced in Section 5.1.

This is the new version; the first two statements are the same as in section 5.1, the third is a modification:

(1) $G = [V, E]$
(2) $V = \{v_1, v_2, \ldots, v_n\}$
(3) $E = \{[v_1, v_3], [v_1, v_6], [v_2, v_n], \ldots, [v_5, v_7], \ldots\}$

The two-element subsets (or vertex-pairs) of our undirected graphs are replaced by ordered pairs. The effect on the drawn version is that we mark the appropriate directions of edges: they become arrows pointing from the first to the second element of the ordered pair. The effect on the name of the object is that it becomes a directed graph or *digraph*.

Sec. 6.2] **Representation of digraphs**

We will look at the ideas behind relations in Chapter 8; it is sufficient for now that you understand the directional nature of many relations, and their invertibility. For any precedence-type relation, such as 'is older than', there is always the inverse case, 'is younger than', and we choose one or the other. We will find that a lot of examples related to sequence and precedence can be considered as digraphs.

Exercise 6.1.1
Draw up the set representation of the 'rabbit eats grass' digraph, which was the second example in section 5.1.

6.2 REPRESENTATION OF DIGRAPHS

We won't rewrite the work we did in Sections 5.1 to 5.5 on *undirected* graphs so that it applies to directed graphs. Clearly many concepts introduced in connection with undirected graphs — vertex degree, isomorphism, paths and so on — can be extended to directed graphs but we should note that some modifications will be required. For example if an edge is incident on a vertex, then it could be directed either towards the vertex or away from it, so there are two kinds of vertex degree, *in-degree* and *out-degree*. Also an apparently connected digraph may not contain a path between a given pair of vertices. In Fig. 6.1 for example, the 'in-degree' of vertex *a* is

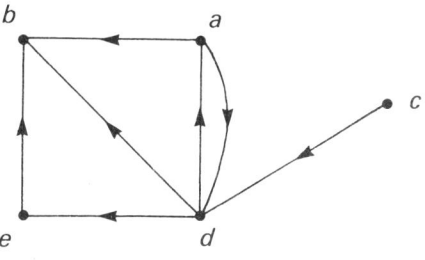

Fig. 6.1.

1 and the 'out-degree' is 2: and we note that there is no path from, say, *b* to *d*, though there are three from *d* to *b*. A directed graph may also contain *loops* — that is, edges starting and ending at the same vertex, like the edge starting and ending at *w* in Fig. 6.2.

An ordered pair is more structured (or ordered!) than a two-element subset (it must be, since two distinct ordered pairs can be lifted from the elements of the one subset), and this extra bit of structure makes the computer handlings of digraphs easier in some ways than that of undirected graphs. For a start, we often represent digraphs in matrix form — and most programming languages have some version of a matrix data type. (Of course our *undirected* graphs could have been represented in matrix form as well, but we chose not to use this representation in Chapter 5.)

We can describe the digraph shown as Fig. 6.1 in terms of a labelled matrix. Note that we read edges *from* a particular row label (in the left-hand column) *to* a particular column label (in the top row) through the element which occupies that row and column in the matrix,

i.e.
$$
\begin{array}{c c} & \begin{array}{c c c c c} a & b & c & d & e \end{array} \\ \begin{array}{c} a \\ b \\ c \\ d \\ e \end{array} & \begin{bmatrix} 0 & 1 & 0 & 1 & 0 \\ 0 & 0 & 0 & 0 & 0 \\ 0 & 0 & 0 & 1 & 0 \\ 1 & 1 & 0 & 0 & 1 \\ 0 & 1 & 0 & 0 & 0 \end{bmatrix} \end{array} \quad \text{or} \quad \begin{array}{c c} & \begin{array}{c c c c c} a & b & c & d & e \end{array} \\ \begin{array}{c} a \\ b \\ c \\ d \\ e \end{array} & \begin{bmatrix} F & T & F & T & F \\ F & F & F & F & F \\ F & F & F & T & F \\ T & T & F & F & T \\ F & T & F & F & F \end{bmatrix} \end{array}
$$

So, for instance, the element in row d, column b of the matrix is a 1 (or a T). This tells us that there is a directed edge in the graph from vertex d to vertex b. In similar fashion row a, column c contains the element 0 (or F), so there is no edge from vertex a to vertex c.

In both of these representations of the graph we are using a Boolean matrix, that is, the 1 and the T are both considered to represent the value TRUE: the 0 and F represent FALSE. Although it would be more consistent of us to use the T/F symbols, which have a clear meaning in Boolean arithmetic, we will use instead the 1/0 symbols, which are much easier to distinguish by eye. But don't forget, they represent TRUE and FALSE; the fact that they look like binary digits doesn't mean that they *are* binary digits in this context. We can go one step further now, and describe the vertex set as a 5 × 1 matrix, rather than hang it on the matrix as a label.

i.e.
$$
V = \begin{bmatrix} a \\ b \\ c \\ d \\ e \end{bmatrix} \quad \text{and} \quad E = \begin{bmatrix} 0 & 1 & 0 & 1 & 0 \\ 0 & 0 & 0 & 0 & 0 \\ 0 & 0 & 0 & 1 & 0 \\ 1 & 0 & 0 & 0 & 1 \\ 0 & 1 & 0 & 0 & 0 \end{bmatrix}
$$

So our description of a digraph turns out to be two matrices, V and E, rather than two sets, and if

$$E[i,j] = 1 \qquad 1 \leq i,j \leq n$$

then there is a directed edge from $V[i]$ to $V[j]$.

Sec. 6.3] Alternative representations

Example 1
Describe the graph which is shown in Fig. 6.2 as a matrix.

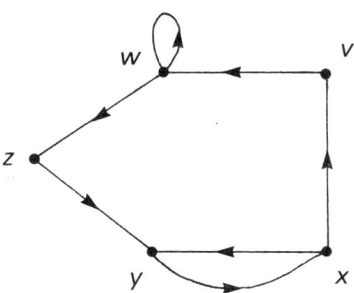

Fig. 6.2.

Answer

$$V = \begin{bmatrix} v \\ w \\ x \\ y \\ z \end{bmatrix} \quad E = \begin{bmatrix} 0 & 1 & 0 & 0 & 0 \\ 0 & 1 & 0 & 0 & 1 \\ 1 & 0 & 0 & 1 & 0 \\ 0 & 0 & 1 & 0 & 0 \\ 0 & 0 & 0 & 1 & 0 \end{bmatrix}$$

Example 2
Draw the graph which is specified in the two matrices V and E.

$$V = \begin{bmatrix} u \\ v \\ w \\ x \\ y \end{bmatrix} \quad E = \begin{bmatrix} 0 & 1 & 0 & 0 & 1 \\ 1 & 0 & 0 & 1 & 0 \\ 0 & 0 & 0 & 1 & 0 \\ 0 & 0 & 1 & 0 & 0 \\ 0 & 0 & 1 & 1 & 0 \end{bmatrix}$$

Answer
A possible drawing is shown in Fig. 6.3.

6.3 ALTERNATIVE REPRESENTATIONS

The two forms of digraph representation considered so far (drawing and matrix) are not the only ones available, and it is worth looking at some alternatives, given that

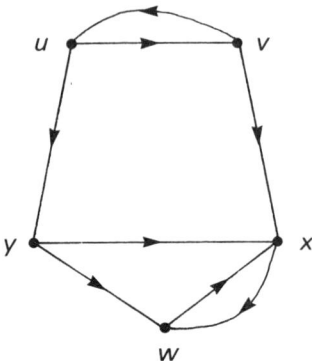

Fig. 6.3.

any representation is likely to have disadvantages, as well as advantages. We will first take a brief look at some characteristics of the two forms we have used up to now.

The set-and-ordered-pair representation of digraphs described in section 6.1 doesn't lead to any convenient implementations; you may recall that the set-and-subset representation of undirected graphs had similar limitations. The matrix representation, on the other hand, is serviceable and — as we will see in section 6.5 — is capable of extension to what are called edge-weighted graphs.

However, there is a problem with matrix representation, namely that for large graphs, which have vertices of fairly low degree, there will be an awful lot of Boolean values equal to FALSE: the matrix, in other words, will be very sparse. For instance a digraph of 100 vertices, with an average vertex degree (in-degree + out-degree) of 5 would have 10 000 entries, of which 9 500 would be Boolean 0's. Fortunately there are techniques for storing sparse matrices efficiently in terms of computer space and access, though we will not pursue the matter here. A further objection to the matrix method is the difficulty involved in modifying the graph: edges are easy to introduce onto or (delete from) existing vertices, as we simply amend entries to the body of the matrix. However, the introduction or deletion of *vertices* is not so easy, as it entails a change in the dimensions of the matrix.

What we will do now is to show how a *linked list* can represent a digraph. In fact this data structure is capable of representing either a directed or an undirected graph — the only difference is that we use two-element sets for edges in the case of a graph, ordered pairs in the case of a digraph, so you can easily amend the following description of you want to. A graph does not have the obvious linear characteristics of a list, so essentially we use the linked list as a straightforward *catalogue* of edges. As such, the representation is not initially very interesting, but remember that the importance of a linked list as a data structure is that it allows the attachment or deletion of elements at any position in the list, whereas a simple list is more restricted, allowing addition or removal only at the tail of the list. (Refer back to Chapter 2 if you need to remind yourself about list operations.)

Sec. 6.3] Alternative representations

A digraph is drawn in Fig. 6.4, and its linked-list representation is given below.

Start	Address	Contents	Pointer
1			
	1	[c,w]	2
	2	[y,b]	3
	3	[w,a]	4
	4	[b,a]	5
	5	[b,y]	6
	6	[b,y]	7
	7	[w,b]	8
	8	[c,b]	0

The list itself looks sequential on paper, but as you will see, we can pick off rows of the list (and hence edges of the graph) in any pre-determined order. Each linked record in this case describes an edge of the digraph. Remember that the pointer tells us the location of the next record, the zero pointer indicating the end of the list. We could draw the linked list itself as a digraph, but this might become confusing, so instead we will keep records in a table.

Each line in the table corresponds to the number (or address) of the record. (If the table were stored on computer, the records could be widely scattered and have much more complex addressing).

After the start pointer, which points to address 1, the pointer on each record simply indicates the record on the next row. The information contained about the graph is very close to the ordered-pair type of description mentioned in section 6.1 — all we are doing is to put the ordered pairs in a list. It wouldn't have mattered in what order we filled in the contents column, we would still be describing the same graph. Here is another representation of the same graph; again the order of edges is immaterial, but we must have a set of pointers and labels which give access to all records.

Start	Address	Contents	Pointer
4			
	1	[c,w]	5
	2	[y,b]	8
	3	[w,a]	6
	4	[b,a]	3
	5	[b,y]	0
	6	[c,y]	2
	7	[w,b]	1
	8	[c,b]	7

In this case the start pointer is to address 4, which gives up edge [b,a]. The pointer

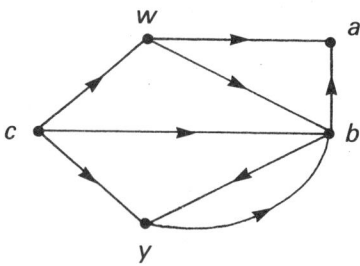

Fig. 6.4.

then directs to address 3, which contains [w,a]. The next pointer directs to address 6, and so on, until the last record reached is at address 5. The zero pointer indicates the end of the list.

This linked-list notation is very straightforward; nevertheless it is worth checking your understanding of it. Check that the links are in a valid sequence: don't just write down the edges!

Example 1
Draw the graph corresponding to the linked list given below

Start 3	Address	Contents	Pointer
	1	[d,e]	4
	2	[e,d]	5
	3	[a,b]	6
	4	[b,c]	3
	5	[a,c]	0
	6	[e,c]	2

Answer
The digraph is shown in Fig. 6.5.

Example 2
Verify that the graph in Fig. 6.6 and the linked list below represent the same information. What is missing 'start' address for the linked list?

Start ?	Address	Contents	Pointer
	1	[d,c]	3
	2	[a,d]	5
	3	[d,b]	6
	4	[c,b]	6
	5	[d,e]	4
	6	[a,b]	2

Answer
The start value for the linked list is 1, and the access sequence is 1, 3, 6, 2, 5, 4.

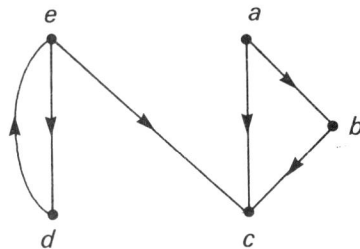

Fig. 6.5.

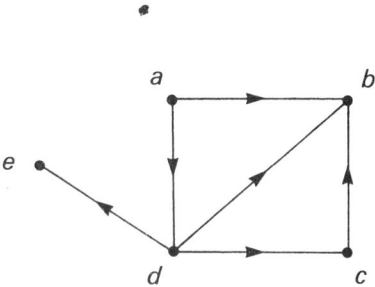

Fig. 6.6.

On first acquaintance the linked-list data structure may appear rather arbitrary — many different lists can represent the same graph — but this arbitrariness makes linked lists very adaptable to circumstance. We will see more examples of this in Chapter 7 (Trees); meanwhile we now examine how a list description might be adapted according to the purpose of the graph. For example, in section 5.4 we looked at a test for connectedness in a graph, which involved considering all vertices adjacent to a given vertex. An efficient set of links in this case might be one which would produce in sequence all the direct successors of a given vertex.

Suppose we take the digraph example given earlier in this section (Fig. 6.4), which we have shown above in two different linked-list representations. In the first representation, the successors of vertex *c* are identified only after a complete pass (eight steps) through the linked list, whereas with the amended set of links in the second representation, they are identified after three steps. (On an example of this size the difference is trivial, but in a large-scale example, such as a database application, it would not be.) So it looks as if we would like to link all graph edges which share the same tail vertex, (the tail vertex being, slightly confusingly, the one at the 'head' of the ordered pair!).

Let's rewrite the edges in alphabetical order of tail vertex (ignore the order of the head vertex). We treat all edges with the same tail as members of the same sublist. The addresses are included so that you can refer to the original linked list — this shuffling of the order is only temporary.

Address	Contents	
4	[b,a]	sub-list 1
5	[b,y]	
1	[c,w]	sub-list 2
8	[c,b]	
6	[c,y]	
3	[w,a]	sub-list 3
7	[w,b]	
2	[y,b]	sub-list 4

Now, with changing the location (or address) of the records in the original table (or for that matter, the order of rows), we can simply change the pointers in order to produce the required linked sub-lists; then we can link the sub-lists to get an efficient linked list sequence, as shown below.

Start	Address	Contents	Pointer
4			
	1	[c,w]	8
	2	[y,b]	0
	3	[w,a]	7
	4	[b,a]	5
	5	[b,y]	1
	6	[c,y]	3
	7	[w,b]	2
	8	[c,b]	6

(You should check that this set of links correctly matches the order of the sub-lists given above.)

Finally we could put an extra level into the hierarchy and have what we might call a 'boss-list', which is a simple list of starting addresses for each sub-list. (This would be particularly useful if the graph is to be amended and updated from time to time.) The pointer at the end of each sub-list is set to 0, which is now interpreted as referring us to the boss-list. The boss-list itself consists of addresses of the first record of each sub-list, and terminates in a 0. This is what the new version looks like.

Sec. 6.4] An application of the matrix form

Boss-list ⟨4,1,3,2,0⟩	Address	Contents	Pointer
	1	[c,w]	8
	2	[y,b]	0
	3	[w,a]	7
	4	[b,a]	5
	5	[b,y]	0
	6	[c,y]	0
	7	[w,b]	0
	8	[c,b]	6

The 4 in the boss-list directs us to address 4, which locates [b,a], followed by a pointer to address 5, which locates [b,y]. The pointer on this record is 0, so we take the next element on the boss-list, which is 1. This locates [c,w] which is the first of the edges which have c as the tail vertex, and so on.

6.4 AN APPLICATION OF THE MATRIX FORM

We'll now look at a possible use of the matrix type of digraph representation in data processing. First we will carry out some simple mathematics on the matrix; after that we will consider how the result might be applied.

We will be using the idea of the logical product of two matrices as explained in Chapter 4, so if you are not reasonably familiar with this idea, then check it before you continue. Suppose we have the following description of the small digraph shown in Fig. 6.7 (don't forget that a 0 represents a Boolean FALSE, and a 1 represents a

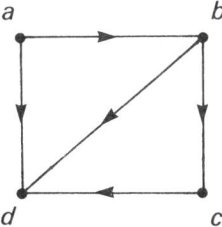

Fig. 6.7.

Boolean TRUE):

$$V = \begin{bmatrix} a \\ b \\ c \\ d \end{bmatrix} \quad E = \begin{bmatrix} 0 & 1 & 0 & 1 \\ 0 & 0 & 1 & 1 \\ 0 & 0 & 0 & 1 \\ 0 & 0 & 0 & 0 \end{bmatrix}.$$

We are going to multiply E by itself — to square it, in fact.

$E \times E$ produces the following

$$
\begin{array}{c} \\ a \\ b \\ c \\ d \end{array}
\begin{array}{c} \\ * \\ \\ \\ \end{array}
\begin{bmatrix} a & b & c & d \\ 0 & 1 & 0 & 1 \\ 0 & 0 & 1 & 1 \\ 0 & 0 & 0 & 1 \\ 0 & 0 & 0 & 0 \end{bmatrix}
\begin{array}{c} * \\ \\ \\ \\ \end{array}
\times
\begin{bmatrix} a & b & \overset{*}{c} & d \\ 0 & 1 & 0 & 1 \\ 0 & 0 & 1 & 1 \\ 0 & 0 & 0 & 1 \\ 0 & 0 & 0 & 0 \\ & & \underset{*}{} & \end{bmatrix}
=
\begin{bmatrix} a & b & c & d \\ 0 & 0 & *1* & 1 \\ 0 & 0 & 0 & 1 \\ 0 & 0 & 0 & 0 \\ 0 & 0 & 0 & 0 \end{bmatrix}
$$

Now suppose we look in a bit more detail at the product of *just one* row/column; for example row *a* and column *c*. The value of this product is 1 (or TRUE) (the row, column and result are marked with asterisks above). What does the result mean, in terms of the graph?

If we set out the steps in more detail the meaning should become clear (remember that it is *possible* to have a directed edge starting and finishing at the same vertex — a loop). The four logical products involved in multiplying row *a* by volume *c* are listed below

$$E[a,a] \wedge E[a,c] = 0 \wedge 0 = 0$$
$$E[a,b] \wedge E[b,c] = 1 \wedge 1 = 1$$
$$E[a,c] \wedge E[c,c] = 0 \wedge 0 = 0$$
$$E[a,d] \wedge E[d,c] = 1 \wedge 0 = 0$$

These are combined using the OR operation, giving the single result $0 \vee 1 \vee 0 \vee 0 = 1$ (or TRUE). In fact what we have done is to test for the existence of all possible paths which

(1) connect vertex *a* to vertex *c* and
(2) have length 2.

The result, 1 (or TRUE) tells us that there is at least one path of length 2 which is directed from *a* to *c*. The *possible* paths are $\langle a,a,c \rangle$, $\langle a,b,c \rangle$, $\langle a,c,c \rangle$ and $\langle a,d,c \rangle$, but, as the right hand sides of the four calculations show, only $\langle a, b, c \rangle$ exists in our graph.

The complete matrix logical product E^2 therefore contains the value TRUE (or 1) in a given row and column $[i,j]$ if there is a path of length 2 directed from vertex *i* to vertex *j* in the associated graph.

We could show similarly that multiplication of the matrix E^2 by the original matrix E will give a matrix which indicates paths of length 3, i.e.

$$
\begin{bmatrix} 0 & 1 & 0 & 1 \\ 0 & 0 & 1 & 1 \\ 0 & 0 & 0 & 1 \\ 0 & 0 & 0 & 0 \end{bmatrix}
\times
\begin{bmatrix} 0 & 0 & 1 & 0 \\ 0 & 0 & 0 & 1 \\ 0 & 0 & 0 & 0 \\ 0 & 0 & 0 & 0 \end{bmatrix}
=
\begin{bmatrix} 0 & 0 & 0 & 1 \\ 0 & 0 & 0 & 0 \\ 0 & 0 & 0 & 0 \\ 0 & 0 & 0 & 0 \end{bmatrix}
$$

$$E \times E^2 = E^3$$

(Notice that this is a case where the order of multiplication doesn't matter: $E^2 \times E$ is also equal to E^3.) The matrix tells us that there is just one path of length 3, and it is from a to d.

If we were to multiply again by E we would get a matrix of zeroes, indicating that there are no paths of length 4. This means that there are no paths longer then 4 either, as any longer path would have to contain a four-step path and $E_n = [0]$ if $n \geqslant 4$.

Example 1
Square the matrix corresponding to the graph shown in Fig. 6.8. Give an interpretration of

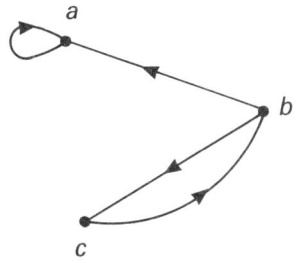

Fig. 6.8.

each occurrence of a 1 (or TRUE) in the squared matrix. If the same edge can be used more than once in a path, what can be said about the longest path in this digraph?

Answer

$$V = \begin{bmatrix} a \\ b \\ c \end{bmatrix} \quad E = \begin{bmatrix} 1 & 0 & 0 \\ 1 & 0 & 1 \\ 0 & 1 & 0 \end{bmatrix} \quad E^2 = \begin{bmatrix} 1 & 0 & 0 \\ 1 & 1 & 0 \\ 1 & 0 & 1 \end{bmatrix}$$

The paths of length 2 are $\langle a,a,a \rangle$, $\langle b,a,a \rangle$, $\langle b,c,b \rangle$, $\langle c,b,a \rangle$ and $\langle c,b,c \rangle$.

We can see from the drawing that there are paths of any length we please. The interesting question is what happens to the matrix for larger powers of E. It will never consist entirely of 1's; for example, there are no paths between b and c which have an even length.

Now we will apply a directed graph and its associated matrix to the development and checking of a simple pay and tax program. This is the starting point for a top-down analysis, so we will only look at the initial outline, but you should see that this method could be extended to any required level of detail. We will consider the processes involved in calculating an employee's after-tax pay, as described by the precedence network shown in Fig. 6.9, and use some basic ideas in graph theory to help analyse the network.

A directed edge from a vertex a, to another vertex b, indicates that the information associated with vertex a is a direct input into the computation of the information corresponding to vertex b. We can describe a as a first-order precedent of b. In a lot of cases a given computation will depend on several first-order precedents. In our example we can see that in order to calculate the gross weekly pay we need three types of information: (1) the grade of the employee, (probably a reference number); (2) details of each grade, such as what weekly salary or hourly rate of pay applies, what the basic hours for that grade are, what the overtime rate is, and so on; and (3) the hours worked by the employee. Notice also that (1) and (2) will usually be the same from week to week, whereas (3) may well vary considerably between employees and between weeks. We will represent the graph of the proposed system in matrix form, and for easier reference we will label the rows and columns of the matrix E.

$$V = \begin{bmatrix} Eg \\ Tc \\ Ra \\ Hr \\ Gp \\ Td \\ Pa \end{bmatrix} = \begin{matrix} \text{Employee's grade} \\ \text{Employee's Tax code} \\ \text{Rate of pay for this grade} \\ \text{Hours worked this week} \\ \text{Gross pay this week} \\ \text{Tax due this week} \\ \text{Pay after tax} \end{matrix}$$

$$E = \begin{matrix} & Eg & Tc & Ra & Hr & Gp & Td & Pa \\ Eg & 0 & 0 & 0 & 0 & 1 & 0 & 0 \\ Tc & 0 & 0 & 0 & 0 & 0 & 1 & 0 \\ Ra & 0 & 0 & 0 & 0 & 1 & 0 & 0 \\ Hr & 0 & 0 & 0 & 0 & 1 & 0 & 0 \\ Gp & 0 & 0 & 0 & 0 & 0 & 1 & 1 \\ Td & 0 & 0 & 0 & 0 & 0 & 0 & 1 \\ Pa & 0 & 0 & 0 & 0 & 0 & 0 & 0 \end{matrix}$$

Each 1 value in the matrix corresponds to one edge in the precedence graph; for example, $E[3, 5] = 1$ means that there is a directed edge from vertex 3 (Ra) to vertex 5 (Gp); or, to go back to the original description, Ra is a first-order precedent of Gp.

We can now consider what other information the rows and columns of this matrix give us. We see for instance, that columns 1, 2, 3 and 4 consist entirely of 0s; this means that Eg, Tc, Ra and Hr have no precedents within the system, and must be provided from outside. In simpler words, Eg, Tc, Ra and Hr are inputs to the system. In the case of rows, there is only one row consisting entirely of 0s; Pa is not a precedent to anything else in the system, it is therefore an output of the system. (You

Sec. 6.4]   **An application of the matrix form**   137

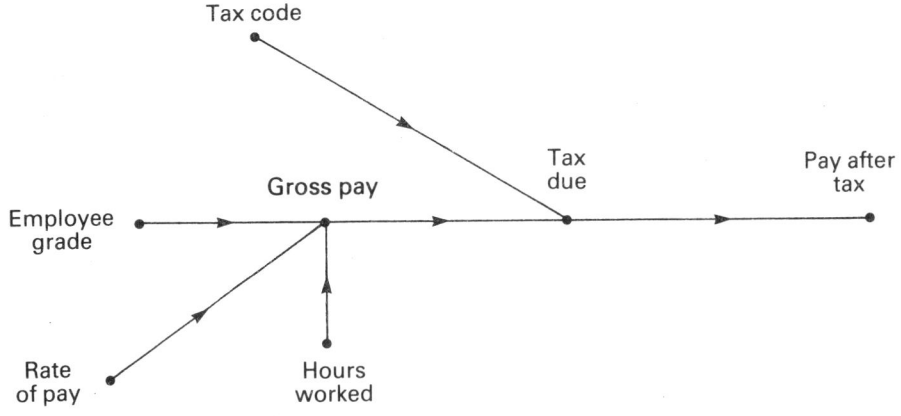

Fig. 6.9.

might ask whether or not Td should also be classified as an output, and if so how we should re-specify the elements of our tax calculation system; one purpose of analysis is to clarify such questions.)

Notice in passing that the main diagonal should consist entirely of 0s; otherwise a piece of information would be its own input, which is clearly impossible.

If we now calculate E^2 we get the following matrix (with labels attached for reference).

$$E^2 = \begin{array}{c} \\ Eg \\ Tc \\ Ra \\ Hr \\ Gp \\ Td \\ Pa \end{array} \begin{array}{c} Eg \quad Tc \quad Ra \quad Hr \quad Gp \quad Td \quad Pa \\ \begin{bmatrix} 0 & 0 & 0 & 0 & 0 & 1 & 1 \\ 0 & 0 & 0 & 0 & 0 & 0 & 1 \\ 0 & 0 & 0 & 0 & 0 & 1 & 1 \\ 0 & 0 & 0 & 0 & 0 & 1 & 1 \\ 0 & 0 & 0 & 0 & 0 & 0 & 1 \\ 0 & 0 & 0 & 0 & 0 & 0 & 0 \\ 0 & 0 & 0 & 0 & 0 & 0 & 0 \end{bmatrix} \end{array}$$

This indicates second-order precedences; for instance Hr is a second-order precedent of Td, since the number of hours worked is a first-order precedent of gross pay, and gross pay is in turn a first-order precedent of tax due. This shows up simply as

$$E^2 [4, 6] = 1$$

Notice that the same item of information could be both a first-order precedent and a second-order precedent to another item. For instance, $E^2 [5, 7] = 1$ and $E [5, 7] = 1$ as well, so that Gp is both a first-order precedent and a second-order precedent of Pa.

Another point to note is that the number of 1 (TRUE) entries in the second-order precedence matrix is eight, compared with seven entries in the original first order matrix E. This means that there are at least eight second-order precedences. (It is possible that there are more than eight — the Boolean value 1 tells us of the *existence* of a path between two specific vertices, but a Boolean variable can't tell us how *many* paths.

We will now consider E^3 and E^4

$$E^3 = \begin{array}{c} \\ Eg \\ Tc \\ Ra \\ Hr \\ Gp \\ Td \\ Pa \end{array} \begin{array}{c} \begin{array}{ccccccc} Eg & Tc & Ra & Hr & Gp & Td & Pa \end{array} \\ \left[\begin{array}{ccccccc} 0 & 0 & 0 & 0 & 0 & 0 & 1 \\ 0 & 0 & 0 & 0 & 0 & 0 & 0 \\ 0 & 0 & 0 & 0 & 0 & 0 & 1 \\ 0 & 0 & 0 & 0 & 1 & 0 & 1 \\ 0 & 0 & 0 & 0 & 0 & 0 & 0 \\ 0 & 0 & 0 & 0 & 0 & 0 & 0 \\ 0 & 0 & 0 & 0 & 0 & 0 & 0 \end{array} \right] \end{array}$$

and

$$E^4 = \begin{array}{c} \\ Eg \\ Tc \\ Ra \\ Hr \\ Gp \\ Td \\ Pa \end{array} \begin{array}{c} \begin{array}{ccccccc} Eg & Tc & Ra & Hr & Gp & Td & Pa \end{array} \\ \left[\begin{array}{ccccccc} 0 & 0 & 0 & 0 & 0 & 0 & 0 \\ 0 & 0 & 0 & 0 & 0 & 0 & 0 \\ 0 & 0 & 0 & 0 & 0 & 0 & 0 \\ 0 & 0 & 0 & 0 & 0 & 0 & 0 \\ 0 & 0 & 0 & 0 & 0 & 0 & 0 \\ 0 & 0 & 0 & 0 & 0 & 0 & 0 \\ 0 & 0 & 0 & 0 & 0 & 0 & 0 \end{array} \right] \end{array}$$

So we see from E^3 and E^4 that there are three precedences of order 3 and none of order 4.

Finally in this small example we will consider a mis-specification in the graph, and, remembering the previous section, will ask what would happen if by error one of the edges had been given the wrong direction — for example if employee grade, had been mistakenly drawn as a successor of gross pay; rather than as a predecessor. The erroneous version of E would look like this.

$$E = \begin{array}{c} \\ \\ Eg \\ Tc \\ Ra \\ Hr \\ Gp \\ Td \\ Pa \end{array} \begin{array}{c} \begin{array}{ccccccc} Eg & Tc & Ra & Hr & Gp & Td & Pa \end{array} \\ \left[\begin{array}{ccccccc} 0 & 0 & 0 & 0 & 0 & 0 & 0 \\ 0 & 0 & 0 & 0 & 0 & 1 & 0 \\ 0 & 0 & 0 & 0 & 1 & 0 & 0 \\ 0 & 0 & 0 & 0 & 1 & 0 & 0 \\ 0 & 0 & 0 & 0 & 0 & 1 & 1 \\ 0 & 0 & 0 & 0 & 0 & 0 & 1 \\ 0 & 0 & 0 & 0 & 0 & 0 & 0 \end{array} \right] \end{array}$$

So employee grade has no successors, according to the first row of the matrix, and would be described by our analysis as an output.

To find the effect of mis-specifying an output (such as *Pa* in this example) you might like to consider the effect of tax due being incorrectly drawn as a successor to pay. You will see that tax due becomes classed as an output.

6.5 EDGE-WEIGHTED GRAPHS

From the start of this chapter we have taken care to define graphs in formal terms of vertices and edges, and when we have drawn the graph we have labelled the picture accordingly. As anyone who's done school geometry will know, labelling is also a simple way of pointing to different parts of the picture, and avoiding ambiguity — we identify a particular vertex in a diagram as x, say, and an edge as $[x, y]$. The existence of labels on graphs is a consequence of our formal definitions of graphs and digraphs; if we hadn't used such definitions there would be nothing for a present-day computer to process. (Computers are good at many tasks; interpreting drawings is not one of them.)

As soon as we used a graph as a model, the labels took on a second meaning: each label corresponded to an element of the object or system being modelled. At various points in this chapter we've had vertices named after George, tax payable, a line in a computer program and so forth.

Now we are going to consider adding extra information to the basic graph and digraph: more precisely we are going to introduce *quantities*, which may be associated with either vertices or edges. As you will see, these extended graphs have a number of applications, and should extend your ideas about modelling. Again, we will use formal (set and matrix) and less formal (picture) representations as appropriate.

One way of extending the scope of a graph is to attach numerical *weights* to each edge. We will name one vertex as a starting point or source s, and another as a finishing point or terminal, t. We will assume that the graph is connected in such a way that there is at least one path from s to t. The graph shown in Fig. 6.10, for example, might represent a network of roads between towns with edge weights

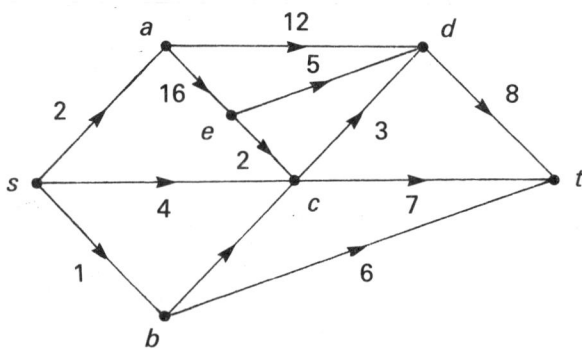

Fig. 6.10.

representing distances (don't worry for now about the one-way-street restriction!), or it might represent connections between stations in a telecommunications network, with unit costs marked on edges. There are three interconnected questions which we are going to look at.

(1) Can we find the path from s to t which has the minimum sum of weights?
(2) Can we devise a general procedure to find the minimum weight path for any directed graph?
(3) Can we store the graph in computer-manageable form, so that we can produce a formal algorithm for the procedure in (2)?

The answer to (1) is clearly yes: we could list all possible paths, and pick out the one with minimum weight. For example

path $\langle s,b,c,t \rangle$ has a cost of $1 + 2 + 7$ ($= 10$)
and $\langle s,c,e,t \rangle$ has a cost of $3 + 5 + 2$ ($= 10$)

and so on. This 'count them all' method is perfectly valid, and quite acceptable in a small example. But there will be problems even with small networks — can we be sure we have listed all paths, and should we really have to carry out so many (almost identical) calculations? In order to show that the answers the questions (2) and (3) are also yes, we will look further at our particular example, and then try to draw some general conclusions from the exercise.

Notice that a 'greedy' method — a method of taking the lowest weight (i.e. cheapest, or shortest distance) edge at each stage — won't necessarily give an overall best answer; the greedy choice of edge $\langle s,a \rangle$, weight 2, forces us to take the edge $\langle a,d \rangle$ with the high weight of 12, and there's worse to come! (There are some valid 'greedy' methods, and you will see two examples in the next chapter, but in the present example the approach clearly doesn't work.)

As an alternative, suppose we work outwards from s and record the shortest path to each vertex by means of a numerical *label* on the vertex. This is the basis of a simple, but ingenious algorithm credited to Dijkstra, which we will now demonstrate, before we discuss it in some detail. In brief, we calculate numerical labels for vertices. The label on a given vertex will represent the minimum distance from the source vertex, s, to the labelled vertex: but this label may be updated (reduced) if a shorter path is subsequently found. The label will become *permanent* when it is certain that it has the lowest value which it can possibly have. This probably sounds more complicated than our earlier idea of counting all paths, but if we approach it systematically, it can be seen that it is easy to apply, and also that unnecessary paths will not be followed once it has become clear that such a path cannot be part of a minimum path.

The table gives the vertex names along the top. Each pair of rows represents an iteration in two stages. The first part, shown in the rows labelled (i) consists in taking the vertex which received a permanent label on the previous iteration, and updating the temporary labels of every vertex directly connected to it. The second part, shown in the rows labelled (ii) consists in the promotion of the smallest temporary label to permanent label, and we indicate permanent status by means of a circle. (You should sketch a copy of the graph, and label the vertices as you work through this tabular description.)

	Vertex						
Iteration	s	a	b	c	d	e	t
0(i)	0	M	M	M	M	M	M
0(ii)	⓪	M	M	M	M	M	M
1(i)	0	2	1	4	M	M	M
1(ii)	⓪	2	①	4	M	M	M
2(i)	0	2	1	3	M	M	7
2(ii)	⓪	②	①	3	M	M	7
3(i)	0	2	1	3	14	18	7
3(ii)	⓪	②	①	③	14	18	7
4(i)	0	2	1	3	6	18	7
4(ii)	⓪	②	①	③	⑥	18	7
5(i)	0	2	1	3	6	18	7
5(ii)	⓪	②	①	③	⑥	18	⑦

Lines 0(i) and 0(ii) are not strictly part of the iteration. Line 0(i) shows the initial labelling of zero on vertex s, and the way in which to begin with we assign an arbitrarily high value M to the remaining vertices (you can think of M as a number so high that no *actual* label can ever have a value as high as M — treat it as a million if you like). Obviously, as soon as a true label is found for any of these vertices, the M will be updated — that is, reduced. (This may seem a curious device, but as we will see later, it is necessary in order to start off the formal algorithm properly.)

Line 0(ii): The label on *s* becomes permanent, as it is the smallest of the seven initial temporary labels.

Line 1(i): The immediate successors of the new (so far the only!) permanently labelled vertex are *a*, *b* and *c*, which take the temporary labels of $0+2$, $0+1$ and $0+4$.

Line 1(ii): The smallest temporary label is the 1 on vertex *b*, so *b* gets a permanent label of 1.

Line 2(i): Checking from newly permanent vertex *b*, we find that the temporary label on *c* is reduced from its value of 4 to a *lower* value of 3 ($=1+2$). Notice that the final vertex *t* gets a temporary label of 7 ($=1+6$).

Line 2(ii): The vertex with the *smallest* temporary label is vertex *a*, with weight 2, so this label on *a* becomes permanent. We don't know if the label of 7 on *t* corresponds to a minimum weight path yet — though we *do* know now that the minimum weight of the path from *s* to *t* must be less than or equal to 7.

Line 3(i): Checking from the new permanently labelled vertex *a*, we find that *d* gets a temporary label of 14 ($=2+12$), and *e* gets a label of 18 ($=2+16$).

Line 3(ii): The vertex with the smallest temporary label is *c*, so *c* gets a permanent label of 3.

Line 4(i): Checking from *c*, we find that *d* gets a *new* reduced temporary label of 6 ($=3+3$). Note that the temporary label on *t* stays at 7, as the new possible label for *t* of $3+7$ (that's the value if we use the edge *c* to *t*) is higher than 7.

Line 4(ii): The smallest-valued temporary label is the label on *d*, so *d* gets a permanent label of value 6.

Line 5(i): Checking from *d*, we see that *t* does not get its temporary label updated to $(6+8)$, so in fact no temporary labels are changed on this iteration.

Line 5(ii): The lowest valued temporary label is on *t*, so *t* gets what we were waiting for — a permanent label. We can say that *minimum weight path* from *s* to *t* has a total weight of 7. Another way of saying this is that the *shortest* path from *s* to *t* has length 7 — or total distance 7 — though we have to remember that it depends on the particular meaning that we give to the terms 'weight', 'length' and 'distance'. In another case, such as our communications network example, the path we have found would be the *cheapest* route from *s* to *t*.

It is worth looking back over this informal algorithm, to be clear about one point in particular, that is, the rule by which labels became permanent. You will see that we accumulated a list of distances in each row of the table, starting with a single zero value allocated to *s*, in row 0. The list was extended, so that temporary labels were introduced at each iteration (row i). On any row, the temporary label with the lowest value indicated that a shortest route to its corresponding vertex had been found (all shorter routes having been accounted for). The value of the label represented the shortest distance from *s*. This label was made permanent (row ii). It is important to note that the order in which vertices received permanent labels was not our choice (though if there had been two equal lowest value labels we could have arbitrarily chosen either). In effect, the algorithm continues its iterations until the target vertex gets a permanent label. We can be certain only that the shortest path will be calculated at worst in $(n-1)$ iterations, where the graph has n vertices altogether.

In this application of the shortest-path algorithm, we have established that the *value* of the shortest path is 7, but what we haven't done is to *identify* this shortest path. By means of a slight modification to the algorithm, we can effectively leave a trail. This is because any vertex which is at the *head* of one or more arrows, and which receives a label (either temporary or permanent), does so because of a previous *permanent* label at the *tail* of one of those arrows. So all we need to do, whenever we update a label, is to indicate the identity of the previous permanently labelled vertex — that is, we leave a kind of 'reverse trail'. This is shown below in the modified version of our table, in which the vertex which led to the creation of a temporary label is shown in brackets next to that label. Notice that there is only one case of a reverse trail being modified — that is, the trail from vertex d.

Iteration	Vertex						
	s	a	b	c	d	e	t
0(i)	0	M	M	M	M	M	M
0(ii)	0	M	M	M	M	M	M
1(i)	0	2 (s)	1 (s)	4 (s)	M	M	M
1(ii)	0	2	1	4	M	M	M
2(i)	0	2	1	3	M	M	7 (b)
2(ii)	0	2	1	3	M	M	7
3(i)	0	2	1	3	14 (a)	18 (a)	7
3(ii)	0	2	1	3	14	18	7
4(i)	0	2	1	3	6 (c)	18	7
4(ii)	0	2	1	3	6	18	7
5(i)	0	2	1	3	6	18	7
5(ii)	0	2	1	3	6	18	7

In this case we see that the final backward reference from t is to b; and from b it is to s. So we get the reverse sequence s, b, t.

Below we give an action diagram description of the shortest path (or minimum weight or minimum distance!) algorithm. Note the following conventions:

(1) A temporary label on a vertex v will be written temp(v).
(2) If and when a temporary label becomes permanent, it will be written perm(v)
(3) A vertex v with a permanent label will have a permanently labelled predecessor, which we will call predecessor(v).

(4) The weight on a directed edge $[v, w]$ will be written distance$[v, w]$.

The action diagram could be summarized as

Entry
⎡ Initialize by setting setting perms(s) to 0 and all temporary labels to M
⎢ Iterate until vertex t receives a permanent label.
⎣ Backtrack to identify the shortest path.
Exit

This is expanded below.

Entry
⎡ perm(s) = 0
⎢ predecessor(s) = null (s has a permanent label and no predecessor)
⎢ ⎡ FOR ALL v
⎢ ⎢ temp(v) = M
⎢ ⎣ END FOR
⎢ (Iterate)
⎢ ⎡ DO
⎢ ⎢ Take latest vertex with a permanent label (note, always vertex v)
⎢ ⎢ ⎡ FOR ALL vertices w, such that $[v, w]$ is an edge of the graph
⎢ ⎢ ⎢ newlabel − perm(v) + distance$[v, w]$
⎢ ⎢ ⎢ ⎡ IF newlabel < temp(w)
⎢ ⎢ ⎢ ⎢ temp(w) = newlabel
⎢ ⎢ ⎢ ⎢ predecessor(w) = v
⎢ ⎢ ⎢ ⎣ ENDIF
⎢ ⎢ ⎣ ENDFOR (all successors of v have now been re-evaluated)
⎢ ⎢ min = M
⎢ ⎢ ⎡ FOR ALL vertices x without a permanent label
⎢ ⎢ ⎢ ⎡ IF temp(x) < min
⎢ ⎢ ⎢ ⎢ min = temp(x)
⎢ ⎢ ⎢ ⎢ v = x
⎢ ⎢ ⎢ ⎣ ENDIF
⎢ ⎢ ⎣ ENDFOR
⎢ ⎢ perm(v) = min (we have a new vertex v with a permanent label)
⎢ ⎣ UNTIL t receives a permanent label
⎢ (Backtrack by making a reverse path in the form of a list: recall that the *head* of a list is a list element)
⎢ path = ⟨t⟩
⎢ ⎡ DO WHILE predecessor (head(path)) ≠ null
⎢ ⎢ path = append(⟨predecessor(head(path))⟩, path)
⎣ ⎣ ENDLOOP
Exit

You should try following through this with our example to see how the algorithm works.

Example 1
Draw the directed graph described by the two matrices V and E below; for ease of reading, the entries without corresponding edges in the graph are left empty. Find the weight of the minimum-weight path from s to t, and identify the path by back-tracking.

Sec. 6.5] Edge-weighted graphs

$$V = \begin{bmatrix} s \\ a \\ b \\ c \\ d \\ e \\ f \\ g \\ h \\ t \end{bmatrix} \quad E = \begin{bmatrix} & s & a & b & c & d & e & f & g & h & t \\ s & & 5 & & & & & 7 & 3 & & \\ a & & & 1 & & & & & & & \\ b & & & & 1 & & & & & & \\ c & & & & & 5 & & & & & \\ d & & & & & & & & & 2 & \\ e & & & & 2 & & & & & 3 & \\ f & & & & & 6 & & & & & \\ g & & & 8 & & & & 5 & & & \\ h & & & & 3 & 2 & & & & & \\ t & & & & & & & & & & \end{bmatrix}$$

Answer
The minimum weight path is ⟨s,g,h,d,t⟩ with weight 12. Note that in the case of two contenders tying for a permanent label — *c* and *f* in this case — you are free to choose either (the other will become permanent on the next iteration).

Example 2
Devise a means of finding the minimum weight path by direct reference to the *matrix*, without drawing the digraph.

A longest-path algorithm

We might expect that Dijkstra's algorithm for finding shortest paths by calculating temporary labels, and then selecting the smallest temporary label to become a new permanent label, could be modified to give an algorithm for finding a *longest* path. In fact the modification is not simple — certainly we cannot just replace 'longest' by 'shortest' and 'largest' by 'smallest'.

To see why this is, try applying the shortest-path algorithm to the graph in Fig. 6.11. This doesn't take long — *a*, *c*, and *t*, being directly connected to *s*, get

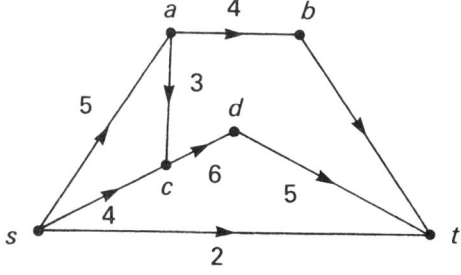

Fig. 6.11.

temporary labels of 5, 4, and 2 respectively. The label of 2 on *t* then becomes permanent, because it is lower than either of the other temporary labels. And as soon as *t* gets a permanent label, the algorithm terminates. None of the other edge-weights can have any effect on the shortest path — roughly speaking, any other route to *t* is bound to be longer than the route of length 2 which we've already identified (since negative edge-weights are not allowed).

Now if we try to find a *longest* path in Fig. 6.12, we can start as before by labelling

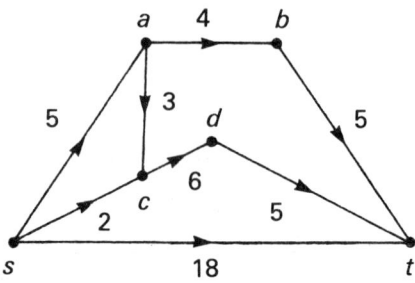

Fig. 6.12.

vertices directly connected to *s* — so *a*, *c* and *t* are labelled 5, 2 and 18 respectively. But we *can't* make the label on *t* permanent this time simply on the grounds that it is the highest figure; it is quite possible — indeed, it is certainly the case here — that examination of one of the other routes to *t* will result in *t* being given a label higher than 18. Until *b* and *d* are labelled, *t* cannot be permanently labelled. In general, we will update a vertex label by looking at the greatest value of all its immediate predecessors, so until those predecessors have been labelled permanently, we cannot update the vertex.

The process, then, is more complex than would at first sight appear. Rather than pursue the formal algorithm for the longest path in an edge-weighted graph, we will instead move on to look at a related, but more useful, concept — that of the *vertex-weighted* graph. But in order to get a little more insight into the problem we've just been looking at, you might like to try the the following first.

Exercise 6.5.1
Verify that in searching for a longest path in Fig. 6.12, the order in which the vertices receive labels, and the value of each label, is as follows:

$[s, 0], [a, 5], [b, 9], [c, 8], [d, 14], [t, 19]$.

(note that the order of *b* and *c* could be reversed.)

6.7 PRECEDENCE GRAPHS REVISITED

The tax calculation example of Section 6.4 required us to order, or to determine precedences for, a number of pieces of information — for example, the information about hours worked preceded the calculation of gross pay, since we could not calculate gross pay until the hours worked were known.

Now we consider a second example which involves the same kind of precedence problems, but this time in relation to a series of activities which together make up the process of receiving and despatching an order. The list of activities, with their estimated durations and the activities immediately preceding them, is as shown:

Activity	Duration (min)	Predecessor(s)
s = send order	—	—
r = receive and process order	30	s
c = check customer details	5	r
a = check goods availability	20	r
d = make despatch arrangements	30	a,c
p = pack goods	40	a,c
t = goods ready for transport	—	d,p

This sequence can be represented by the graph shown in Fig. 6.13, in which vertices

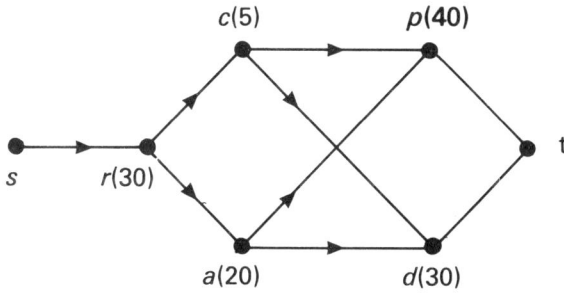

Fig. 6.13.

represent activities; the graph is often called a *network* in this context. Unlike our previous weighted graphs, which were edge-weighted, this one is *vertex*-weighted, the weights representing the durations of the activities. The existence of an edge directed from activity x to activity y indicates that the tail event x must precede the head event y. Notice that we have added a starting vertex s, and that in this case t, the terminating vertex, is already included (in other cases this too might need to be added to the original activity list).

148   **Digraphs**   [Ch. 6

(It is perhaps worth mentioning that there is an alternative way of drawing the network for a sequence of activities, in which the edges rather than the vertices represent activities. This version, often called an arrow diagram, is described in many textbooks on operational research, but it has several technical disadvantages over our 'activity = vertex' model, so we won't pursue it here.)

One question of interest to the person carrying out the order-processing is the (shortest) time taken from the point where an order is received to the stage where goods are ready for transport. This shortest time will be given by the *longest* path through the graph — the one having the greatest sum of vertex weights. This may seem odd at first, but actually it is common sense — there may be some sequences of tasks which can be completed quickly, but the minimum time for the overall project is going to be determined by those activities which take a long time to complete.

We therefore need an algorithm for determining the longest path. Initially, each vertex (apart from the start and finish vertices) will have two labels — an identifier for the activity (a, c, etc.) plus a duration for that activity. The algorithm, like the one for shortest paths discussed earlier, involves giving the vertices two further labels — initially temporary, but subsequently, perhaps after several updates, becoming permanent. These labels represent the earliest starting and finishing times for the activity. We will not give a formal version of the algorithm, just an informal description.

To see how the process works, let's apply it to the graph in Fig. 6.13; you will need to refer to this graph throughout your reading of the next few paragraphs in order to follow what's going on, which is more complicated to explain than to actually do.

We begin by giving the 'dummy' vertex s (= start) a duration of zero; then its earliest start and finish labels also become zero, and can immediately be made permanent.

Now we examine the vertices immediately following s — in this case just r. We give r a temporary start time label of 0 (determined by the permanent finish label for s). As there are no other edges leading into r, and thus no possibility that the start time on r may need updating, this start label can be made permanent. The (permanent) finish label for r then becomes $0 + 30 = 30$.

We next move on to vertices imediately following r — that is, a and c. Both of these are given temporary start labels of 30, determined by the permanent finish label for r. As there are no other edges leading into either of these vertices, both these start labels become permanent, leading to permanent finish labels of $30 + 5 = 35$ for c and $30 + 20 = 50$ for a.

Temporary start labels of 50 now go on d and p, which follow a. We can't make either of these permanent until we have checked the finish label on c, which precedes d and p. As it happens, this value is 35 which is smaller than 50, so the 50s do not change, and can become permanent.

This takes a long time to describe, but can be summed up very simply: the permanent start label on a vertex is equal to the *greatest* of the permanent finish labels on all immediately preceding activities. In plain English, this means that you can't start an activity until all the preceding activities have been finished.

Using this idea, you should now be able to complete the labelling of Fig. 6.13. You should find that t, the terminal vertex, receives a permanent start label of 90, which represents the overall time for the entire project. The fully labelled version is shown in Fig. 6.14.

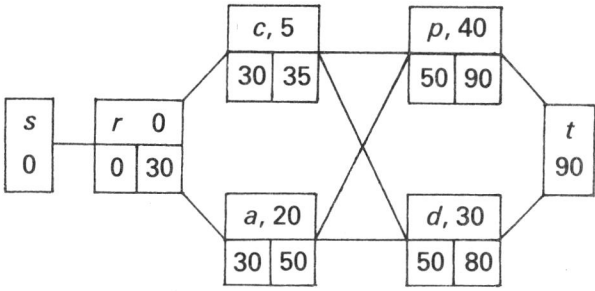

Fig. 6.14.

Once we have the earliest start and finish labels for all the activities, we could add two further labels to each vertex, representing the *latest* times at which those activities can start and finish. For example, vertex d could have a finish time of 90 rather than the present 80, without affecting the start label of 90 on t; so we could give d a latest finish label of 90. That in turn enables us to see that there is 10 minutes spare time — the technical name for this is *float* — on activity d. We will not go through the calculation of the latest times in detail. Fig. 6.14 could be extended to include both earliest and latest times, plus floats; you may like to try this for yourself.

PROBLEMS

1. (a) Draw the digraph represented by the matrix form:

$$G = [V, E]$$
$$V = \{a\ b\ c\ d\}$$

$$E = \begin{bmatrix} 0 & 1 & 0 & 1 \\ 0 & 1 & 1 & 1 \\ 0 & 0 & 0 & 0 \\ 0 & 0 & 1 & 0 \end{bmatrix}$$

(b) Write out a set/ordered pair description of G (with the tail vertex as the first element of each ordered pair, and the head vertex as the second).

(c) Produce a linked-list description of G. Then produce an efficient linking (using an additional 'start' list), so that all edges with vertex a as head are accessed first; then all edges with vertex b as head, and so on.

(d) Suppose it is necessary to append a new edge, $[c, a]$ to the graph. What amendments

are necessary to the four representations (picture, matrix, set/ordered pair, linked list) in order to do this?
(e) Suppose we have to append a new vertex e, and a new edge $[c,e]$; how flexible are our four representations in that case?

2. The Boolean matrix below represents a directed graph.

$$\begin{array}{c} \\ a \\ b \\ c \\ d \end{array} \begin{array}{c} \begin{array}{cccc} a & b & c & d \end{array} \\ \left[\begin{array}{cccc} 0 & 1 & 0 & 1 \\ 0 & 0 & 0 & 1 \\ 0 & 0 & 0 & 0 \\ 0 & 0 & 1 & 0 \end{array} \right] \end{array}$$

(a) Draw the graph.
(b) By an appropriate matrix multiplication, find which pairs of vertices are connected by paths of length 2.
(c) Explain by means of an example how the multiplication produces the required result.
(d) By means of further multiplication, show that there is just one pair of vertices connected by an edge of length 3.
(e) Why would a further matrix multiplication show that there are no paths of length greater than 3?
(f) Describe the effect on your answers of a loop on vertex b.

3. A weighted directed graph is represented by the matrix below.

$$\begin{array}{c} \\ a \\ b \\ c \\ d \\ e \\ f \end{array} \begin{array}{c} \begin{array}{cccccc} a & b & c & d & e & f \end{array} \\ \left[\begin{array}{cccccc} & 2 & & 9 & & \\ & & 5 & & 14 & \\ & & & 1 & & \\ & & & & 7 & \\ & & & & & 2 \\ & & & & & \end{array} \right] \end{array}$$

(a) Describe an appropriate labelling algorithm to find the shortest (minimum weight) path between two vertices in a directed graph — assuming that such a path exists.
(b) Apply the algorithm to the given graph in order to find the shortest path from a to f. Your results should be in the form of a table.
(c) Explain briefly how a *longest-path* algorithm differs from the shortest path algorithm.

4. An auditing system operates as follows. There are six files (or 'books' or ledgers): the Cash_file, Bank_statements, the Petty_cash file, Day_books, the Nominal_ledger and Sales_purchase_ledgers. Each book, file, or ledger may be used on only one job at a time. The jobs are 'activities' and their durations are shown in the table below. Jobs marked * must all be finished before any of the remainder can be done.
(a) Produce a precedence-graph description of the audit, with the jobs a to g as vertices.
(b) This audit is carried out once a week, starting at 10 o'clock in the evening. At what time will it finish? Is any file/ledger/book in continuous use through the audit?

	Activity	Estimated duration (min)
a	Call over Cash_file to Bank_statements	25*
b	Check postings from Cash_file to Petty_cash	17*
c	Test Invoices to Day_books	33*
d	Check postings from Day_books to Sales_purchase_ledgers	30*
e	Check postings from Petty_cash to Nominal_ledger	18
f	Test Cash_file with Sales_purchase_ledgers	25
g	Check postings from Cash_file to Nominal Ledger	28

7
Trees

7.1 TREES

In this chapter we are going to consider a versatile and useful mathematical structure called a tree. In common with its natural namesake, it exists in many forms and can be looked at from several different perspectives, but we hope to show that we can characterize trees into a small number of different types. We will start with the example shown in Fig. 7.1.

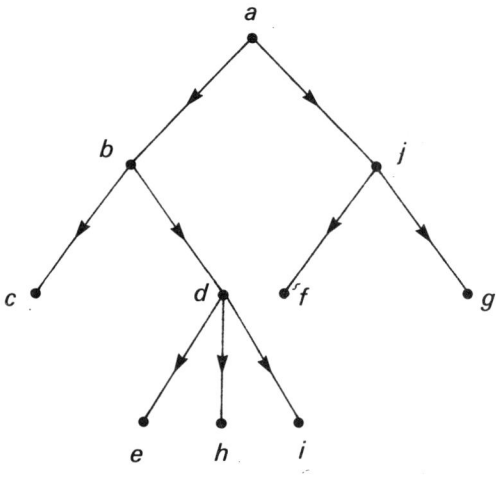

Fig. 7.1.

The picture is of a tree with vertex names $a, b, c \ldots j$. As you can see, in the terminology of Chapter 6 we have a special kind of graph, that is, a connected directed graph with no loops or circuits (this is one possible definition of a tree). It is often useful to specify one vertex as the *root* of the tree. In terms of directed graphs, the root is the only vertex without an edge directed towards it.

Whatever the botanical objections, mathematicians conventionally draw trees in such a way that the root is the highest vertex on the page, and all edges (or branches) are drawn so that any path from the root follows a direction down the page. In our example the root is vertex *a*. To continue the rather shaky biological analogy, all vertices of degree one (other than the root, which could just possibly be called a vertex of degree one) are usually called *leaves*: vertices *c, e, h, i, f* and *g* are leaves in this example. One final introductory definition takes its cue from family trees: if two vertices are connected by an edge, the vertex at the tail end of the edge is known as the *parent* of the other vertex: and the other vertex is known as the *child*. Where a parent vertex has more than one child, the children are sometimes known by the non-sexist name of *siblings*.

Because of the top-down convention for drawing trees, you will find that the edges of the tree are not always marked with a direction, so, although we have put the appropriate arrows onto the edges of Fig. 7.1, we omit them in the remaining examples, and we have assumed a 'down-the-page' direction for all the edges of a given tree. We will say something about the representation of un-rooted trees in Section 7.2. The *spanning trees* which are derived from undirected graphs (see Sections 7.6 and 7.7 below) are unrooted as well. However, unless you are told otherwise, it is usual to treat trees as being rooted, with the root at the top.

Example 1
Verify that the (rooted) tree shown in Fig. 7.2 is isomorphic to the tree shown in Fig. 7.1; that is, all the parent/child/sibling relations are the same; and the root and leaves are the same.

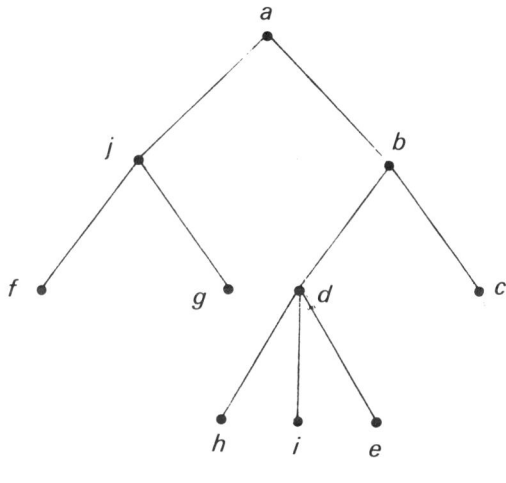

Fig. 7.2.

Example 2
Look at the tree in Fig. 7.3; name the root and the leaves, also list the sets of siblings.

Answer
The root is vertex *v*. The leaves are *r, s, y, z, p* and *q*. The sets of siblings are $\{w,x\}$, $\{r,s\}$, $\{y,z,v\}$ and $\{p,q\}$.

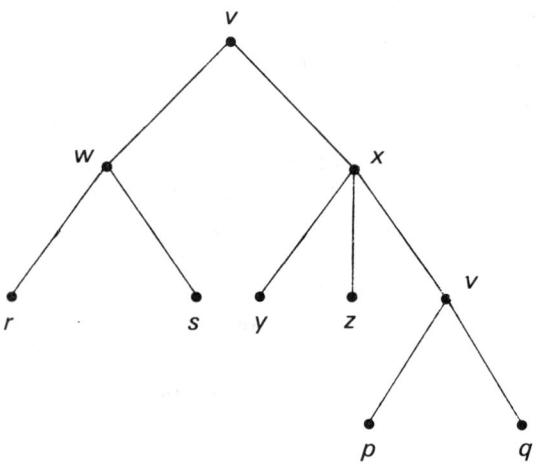

Fig. 7.3.

Example 3
Re-draw the tree shown in Fig. 7.1 (by re-directing the edges, and setting out the shape differently) in order to give vertex b the status of root. Name the leaves and sets of siblings.

Answer
The re-drawn tree is shown in Fig. 7.4. The leaves are c, e, h, i, f and g (as before), and the siblings are $\{c,d,a\}$, $\{e,h,i\}$ and $\{f,g\}$ (note that j is not a sibling of e, h, and i).

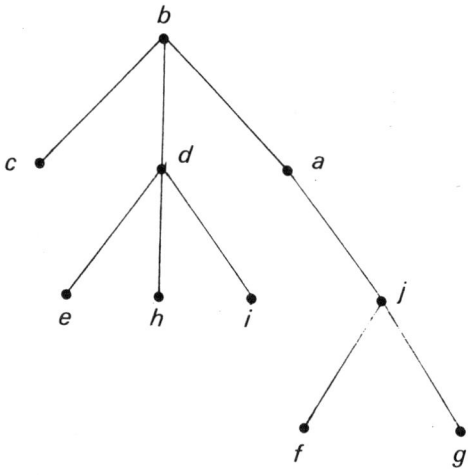

Fig. 7.4.

As you can see, we can use *rooted* trees to represent information which has some kind of *hierarchical* structure — obvious examples are family trees, the departmental structure of a firm, and so forth. Two further examples are shown in Figs 7.5 and 7.6.

These represent respectively a sub-section of the Dewey book index, and the departmental structure of Coketown Polytechnic: these two trees are identical in structure to our first tree in Fig. 7.1, although the information contained is quite different.

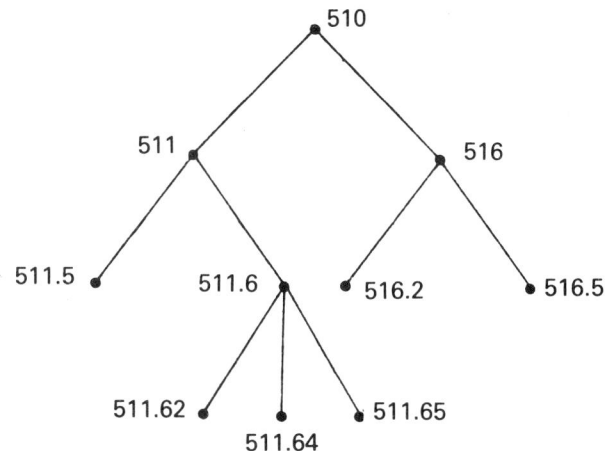

Fig. 7.5.

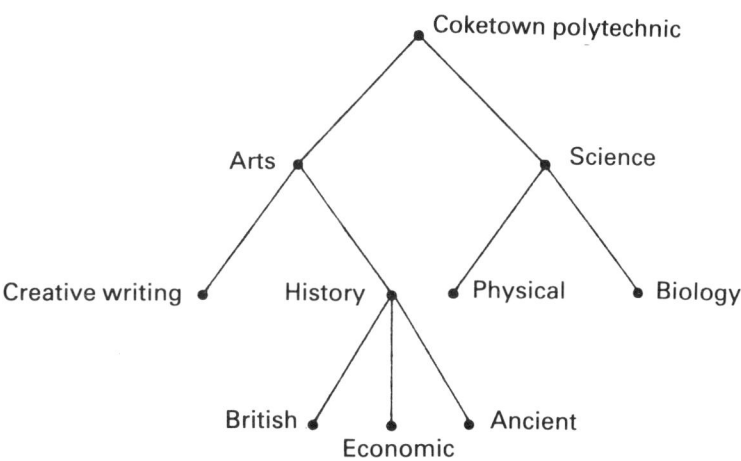

Fig. 7.6.

Apart from this structural similarity of different trees, there is also the matter of when two or more trees can be regarded as identical. Suppose we look back at the trees shown in Figs 7.1 and 7.2. In example 1 above we saw that they both fitted the same description, and now we'd like to ask whether they can be considered to be the same tree. The answer is that it depends: if the left to right *ordering* of siblings matters, then the two trees are not the same. So the next question relates to the information contained by the trees — does the order matter? Given the abstract nature of the trees in Figs 7.1 and 7.2 we can't answer the question for those trees with any certainty. The Dewey index of Fig. 7.5 is best represented as an ordered tree: for example we would expect the siblings 511 and 516 to be in numerical order from left to right, as are all the other sibling groups. In the case of Coketown Polytechnic (Fig. 7.6) there isn't any obvious ordering principle for the tree, either alphabetic or numeric. This ordered/unordered question occurred in connection with ordered pairs and two-element subsets in Chapters 2, 5 and 6, and we'll meet it again in Chapter 9 (methods of counting). (You might note that — as it happens — Figs 7.1, 7.5 and 7.6 have identical ordered structures, and, if we ignore order, Fig. 7.2 has the same structure as the other three: the four trees are isomorphic in the manner of our isomorphic graphs in Section 5.3.) So we need to be aware that in certain circumstances we will be dealing with *ordered* trees; and that in other cases the left to right ordering will not be important. In the next section we will examine a number of ways of representing trees, and you will see that in some cases an ordered tree can be represented by another data structure, and then re-drawn from this representation with the order preserved, but in other cases the order of the original tree is lost (although of course, the parent/child relations are still correct).

7.2 REPRESENTATION OF TREES

We have plenty of choice in the matter of representation of trees. All the representations we used for graphs and digraphs in Chapters 5 and 6 are applicable; and because of the additional conditions and restrictions associated with trees, we can produce a number of other data structures which are equivalent to trees. A lot depends on the particular use to which the data will be put — for example, computers accept information sequentially as a list or string, whereas users of English tend to process information from left to right and down the page, and it's worth being aware of such differences. The first representation — shown in Fig. 7.7 — is obtained by rotating the tree through 90°. The root is clearly indicated; and as the diagram now stands (or has fallen), the tree is still an ordered tree — the left to right ordering of Fig. 7.1 is replaced by a matching top to bottom ordering in this new version.

We won't enter into a discussion of the advantages and disadvantages of this form as far as cognitive psychology is concerned; but it is worth noting that the horizontal form enables us to represent very wide trees more easily. If you think of our normal top-down tree, and imagine that each parent vertex has, say, five children, you will see that the second level of the tree will have a width of 25 vertices (or more accurately 25 vertex *names*), which could easily exceed the width of the page. On the orher hand, our new representation will stretch vertically down the page for 25 lines and should be easy to accommodate. Even if the picture *does* extend beyond the bottom of the page or screen, it's easier to extend drawings vertically — both on paper and on a screen — than it is to extend them sideways.

This horizontal representation gives a clue to another formulation, which has the

Sec. 7.2] Representation of trees

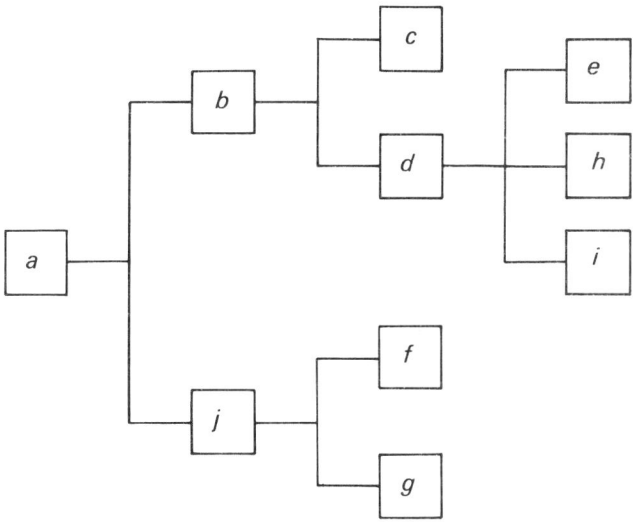

Fig. 7.7.

form of an *action diagram*. What we do is to redraw the horizontal tree according to the following simple rules.

(a) Each parent vertex is associated with a square bracket, whose top edge is marked with the vertex name.
(b) All children are placed in the appropriate parent bracket immediately to the right of the bracket.

This diagram is easy to draw, and is shown in Fig. 7.8. Notice particularly the way in

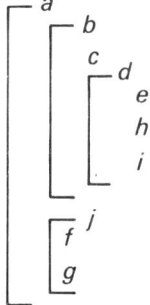

Fig. 7.8.

which the brackets correspond to different levels (or path lengths from the root). Lastly — notice that this representation preserves the *order* of the original tree.

Example 1
Show the horizontal tree form of the Dewey catalogue, followed by its action diagram form.

Answer
See Figs 7.9 and 7.10.

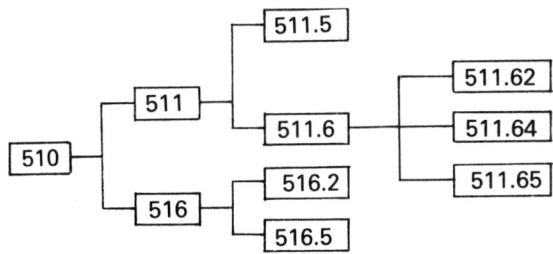

Fig. 7.9.

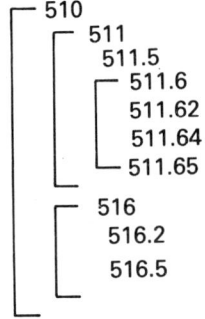

Fig. 7.10.

Example 2
Draw an action diagram representation of the faculty structure of Coketown Polytechnic. Are there possible ambiguities for the casual reader?

Answer
See Fig. 7.11.

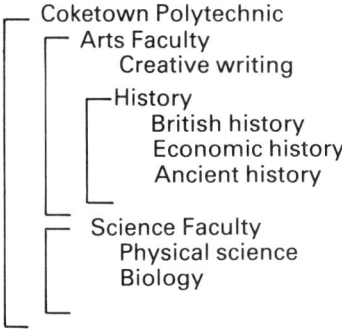

Fig. 7.11.

Example 3
Under what circumstances does the tree drawn in Fig. 7.2 correspond to the tree which is represented as an action diagram in Fig. 7.8?

Answer
The trees correspond if we consider them as unordered trees: as ordered trees they are different.

Since a *non-rooted* tree is a particular form of undirected graph (connected, with no loops or circuits) we could represent the non-rooted version of the tree in Fig. 7.1 as if it were an undirected graph.

$T=[V,E]$
$V=\{a,b,c,d,e,f,g,h,i,j\}$
$E=\{\{a,b\},\{a,j\},\{b,c\},\{b,d\},\{d,h\},\{d,e\},\{d,i\},\{j,f\},\{j,g\}\}$

As a data structure this representation has little to recommend it; it does not make clear the properties which make trees interesting and useful.

However, if we use an ordered-pair representation in the way we used it for digraphs, then we can at least indicate parent/child relations. We can give each edge a direction, say from parent vertex to child vertex. You should be able to identify the root of the tree in this representation.

$T=[V,E]$
$V=\{a,b,c,d,e,f,g,h,i,j\}$
$E=\{[a,b],[a,j],[b,c],[b,d],[d,h],[d,e],[d,i],[j,f],[j,g]\}$

The root is the vertex which has no parent, and therefore its name does not appear as the second element in any of the ordered pairs of E. We can identify leaves of the tree in a similar way, since they will not appear as the *first* element in any ordered pair.

Neither of the representations of a tree we've looked at so far enables us to reconstruct the *ordered* tree. In order to do that, we first need to note a few general points. Point one is that any vertex of a tree may be considered as the root of a sub-tree — the sub-tree being, without getting too technical, the tree consisting of the vertex plus all its descendents, together with their connecting edges. Point two is that, strange though it may seem, a tree can consist of a single vertex and no edges. And point three is that any tree can be identified by its root vertex, so we can refer to 'the tree with root a' rather than describing the complete tree.

Using these ideas, we can build up a way of representing the ordered tree, using list notation rather than set notation, since a list has an implication of order which a set does not. We will use the tree in Fig. 7.1 as an example.

Start at the root, with the simple description of the tree as $T=\langle a \rangle$: T is the tree with root a. However, a has two children, each of which is also a tree — namely left child $\langle b \rangle$ and right child $\langle j \rangle$. We extend our representation to include these thus:

$$T=\langle a \langle b \rangle \langle j \rangle \rangle.$$

Notice that because the list notation implies ordering, the sequence $\langle b \rangle \langle j \rangle$ models the left–right order of the vertices.

We continue in this way, expanding the description of the sub-trees at each stage. The next stage, when the children of b and j are included, is

$$T=\langle a \langle b \langle c \rangle \langle d \rangle \rangle \langle j \langle f \rangle \langle g \rangle \rangle \rangle.$$

Finally we have the description which includes all the vertices explicitly:

$$T=\langle a \langle b \langle c \rangle \langle d \langle e \rangle \langle h \rangle \langle i \rangle \rangle \rangle \langle j \langle f \rangle \langle g \rangle \rangle \rangle.$$

Always check that you have matching numbers of opening and closing brackets in your representation — it is easy to miss one! It is interesting to note that in this representation, anything within a matched pair of brackets is a tree, and the element directly to the right of an opening bracket is the root of that tree.

Exercise 7.2.1
Compare the list representation of Fig. 7.1 obtained above with the action diagram in Fig. 7.8.

Exercise 2.2.2
Give a series of simple rules to enable someone to construct the graphical version of the tree (a picture, in other words), from a single left-to-right reading of the list. For example, $b \langle c$ indicates that c is the first (left-most) child of b; and the symbol $\rangle$ indicates that you should go up one level, to the appropriate parent vertex.

7.3 BINARY POSITIONAL TREES

You may wonder why we need to introduce *another* kind of tree; the answer is that *binary positional trees* combine mathematical simplicity with some quite interesting and flexible applications. Unfortunately the *name* binary positional tree is not

Sec. 7.3] **Binary positional trees**

universally established, and you may find other titles used: binary ordered tree, or positional binary tree, for example. You should always check which kind of tree is being referred to. In this section we'll start by describing this type of tree.

Firstly it is an *ordered* tree; secondly, any vertex has at most *two* children (hence the name binary). The new idea relates to these children. In the case of a vertex with no children, we consider the vertex to be a leaf, as we have done before. But if a vertex has two children, then we have to distinguish between them in a particular way. With an ordered tree, we would list a group of siblings numerically as the first child, second child, third child and so on, reading from left to right. With a binary positional tree, however, we name the children as *left* child and *right* child. This new form of naming probably appears unnecessary, but if we consider the case of a vertex with just *one* child we see that there *is* a distinction. The word 'second' implies a first, but a left can exist without a right — just as a right can exist without a left. So in this one sense, a binary positional tree carries extra information: a single child may be specified — and drawn — either as a *left* child or as a *right* child, and the distinction is important. An example of a binary positional tree is shown in Fig. 7.12.

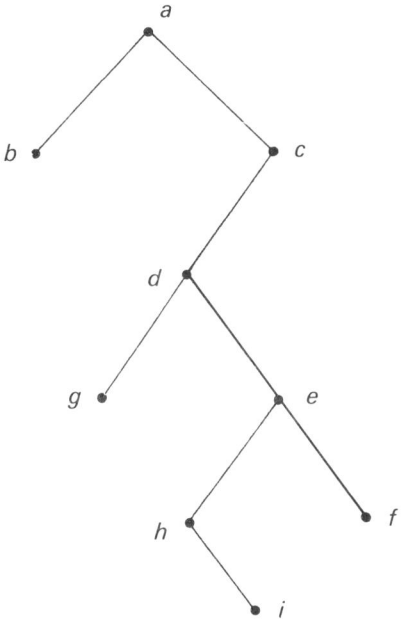

Fig. 7.12.

Example
For the binary positional tree in Fig. 7.12 name the vertices which have (a) two children, (b) a left child only and (c) a right child only.

Answer
(a) *a, d, e* (b) *c* (c) *h*.

Linked-list representation

Now we'll try to show what's special about binary positional trees. Firstly — as we are going to show in Section 7.5 — it is always possible to represent the information carried by an *ordered* tree in the form of a corresponding binary positional tree (and to restore the ordered tree should we so wish). Secondly there are just four kinds of vertex: those with *no* children; those with a *left* child but no right child; those with a *right* child but no left child; and those with *two* children — one labelled left and the other labelled right. This feature makes it very easy to map a binary positional tree onto a computer-useable data structure called a *doubly linked list*.

If you recall the use of a *singly* linked lists in Section 6.3 to represent a digraph, you will remember that a record (x, y) stored at address j represented the existence of the directed edge (x, y) in the digraph, and that a single pointer pointed to the address of another record. The doubly linked list corresponds to the binary positional tree in the following way.

The *record* in the doubly linked list corresponds to a *vertex* in the tree, and there are *two* pointers, which we will call the left and right pointers respectively. The left pointer gives the address of the left child: and the right pointer gives the address of the right child. A zero (or null) pointer means that the corresponding child does not exist. It's not clear which came first, doubly linked lists or binary positional trees; but it is the case that the two structures match admirably: the only feature of the tree which the doubly linked list fails to capture directly is the address of a given vertex's parent, but in programming terms this is no problem. An example of a binary positional tree and its doubly linked list form is shown in Fig. 7.13.

A mapping of an ordered tree onto a binary positional tree

We are now going to show that *any* tree can be conveniently represented, and stored on a computer, in the form of a doubly linked list. This is a consequence of the following series of facts.

(1) An unordered tree may be represented in one of its ordered forms without any loss of information.
(2) An ordered tree may be represented by an equivalent binary positional tree.
(3) A binary positional tree may be represented as a doubly linked list.

(1) and (3) here have already been established. It remains to show that (2) is true. We will show one way of mapping an ordered tree onto a binary positional tree, and also provide an inverse mapping so that the ordered tree can be recovered from the binary positional tree.

Suppose we refer to the ordered tree as T and the binary positional tree as B. Although both T and B will contain the same vertices, the relation between the vertices will differ according to which tree we are considering. A vertex, when considered as a member of T may have zero, or one, or several children; and we would list the several children of a given vertex as first, second, third and so on (that is what we mean by an ordered tree).

Binary positional trees

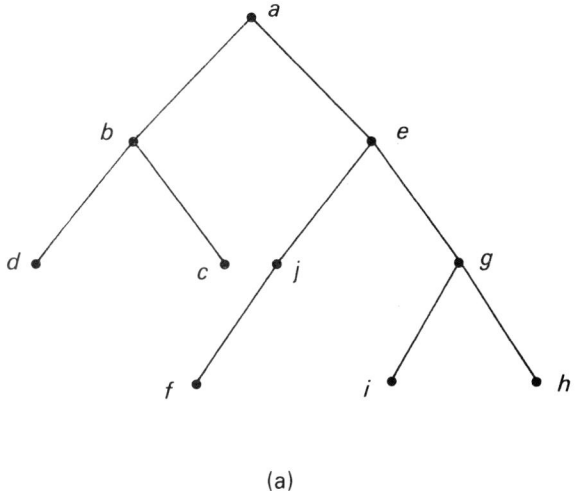

(a)

Address	Record	L. child	R. child
1	a	3	5
2	d	0	0
3	b	2	4
4	c	0	0
5	e	10	7
6	f	0	0
7	g	9	8
8	h	0	0
9	i	0	0
10	j	6	0

Fig. 7.13.

A vertex of B, on the other hand, may have zero, one or two children. In the case of a vertex with one child, the child must be named as either left child or the right child of the vertex. In the case of a vertex with two children, one child must be specified as the left child of the vertex, and the other as the right child. If this much is clear to you, then the correspondence between B and T is easily understood, and the conversion can be carried out either from T to B, or vice versa. Notice that in the description below, v is any specific vertex, and we are describing the status of v in both B and T

(1) The left child of v in B corresponds to the first child of v in T.
(2) The right child of v in B corresponds to the next *sibling* of v in T.

In simple terms this means that to convert from T to B, the first child of vertex v in T becomes the left child of v in B. The next sibling (if it exists) of v in T becomes the right child, and the next sibling becomes the right child of that right child, and so on. Fig. 7.14 should make this clear. Note that the root vertex r in T has three children a, b and c, whereas the root vertex r in B has just one child, a, which is a left child (the root of any binary positional tree thus obtained always has just one child, which is always a left child). The vertices b and c in T become the right child and grandchild in the binary positional version.

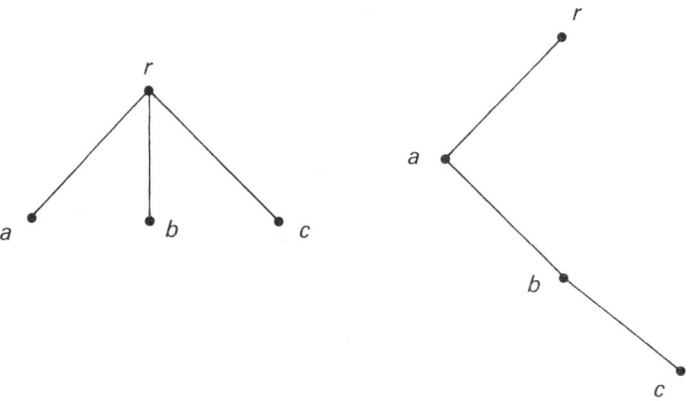

Fig. 7.14.

There are other possible conversions, as you might expect, but this version is as good as any, and with very little practice becomes easy. Notice that an ordered tree T may look like a binary positional tree, nevertheless its binary positional representation B is quite different. Example 2 below will make this clear.

Example 1
Draw and compare the binary positional representations of the two ordered trees T_1 and T_2 shown in Fig. 7.15(a) and (b).

Answer
See Fig. 7.16.

Example 2
Look back at the trees in Figs 7.1 and 7.2. Convert them into binary positional trees.

Sec. 7.3] **Binary positional trees** 165

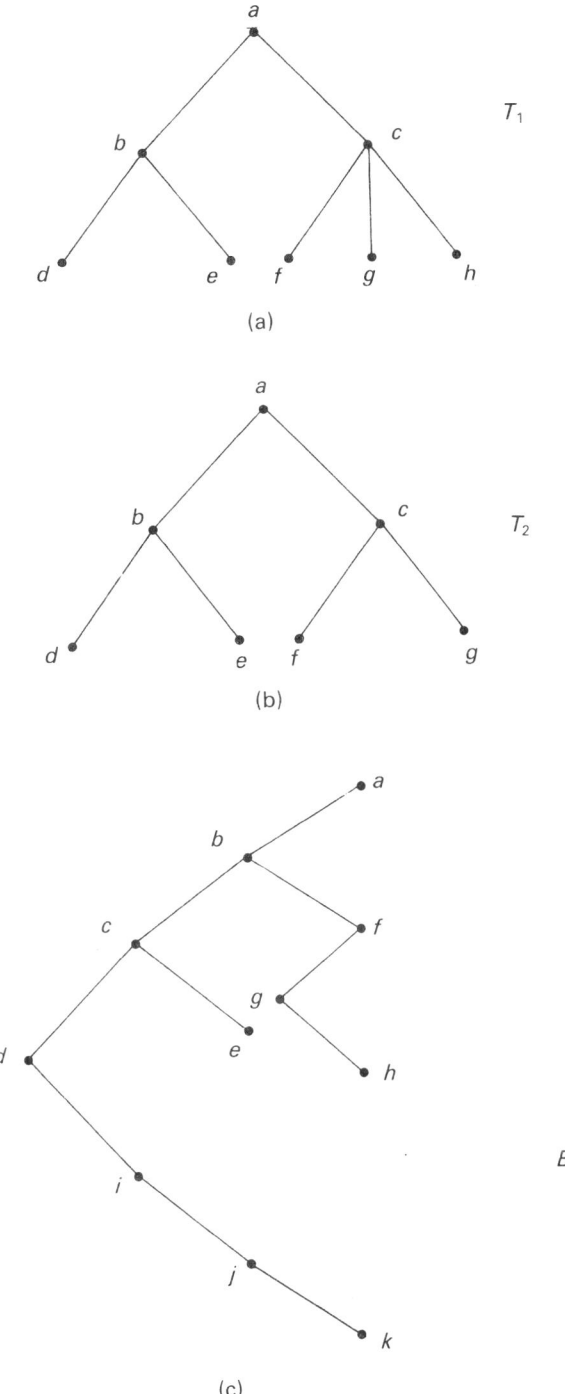

Fig. 7.15.

166 Trees [Ch. 7

Answer
See Fig. 7.17.

Example 3
Convert the binary positional representation B_3 (Fig. 7.15(c)) into an ordered tree form.

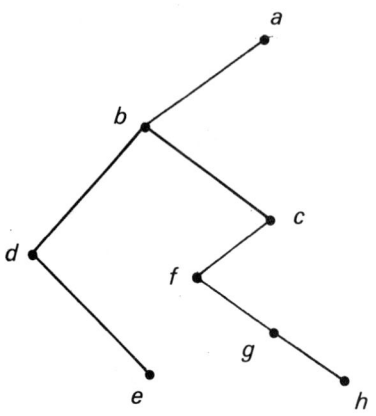

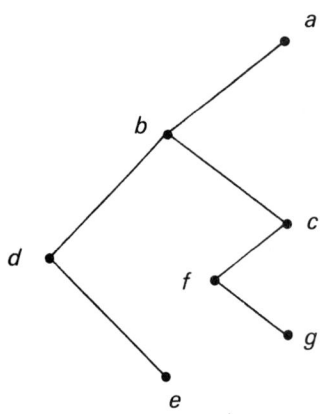

Fig. 7.16.

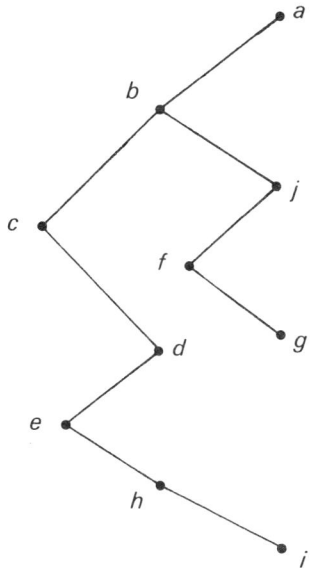

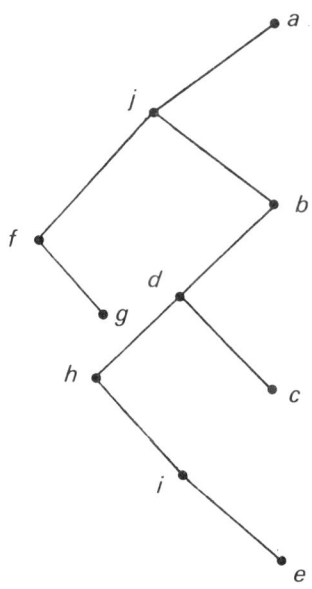

Fig. 7.17.

168   Trees   [Ch. 7

Answer
See Fig. 7.18.

Example 4
Convert the binary positional tree given as the linked list L_4 (Fig. 7.19) into an ordered tree.

Answer
See Fig. 7.20.

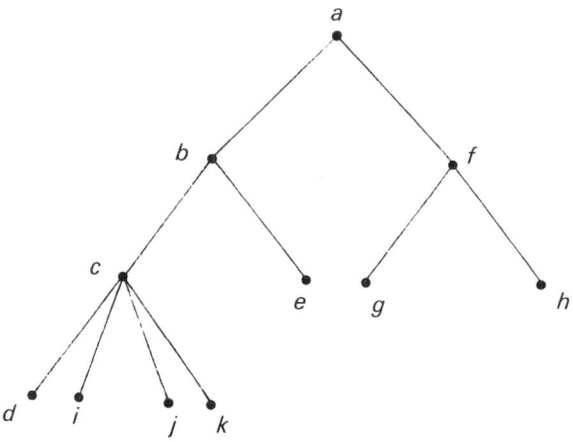

Fig. 7.18.

Address	Record	L. child	R. child
1	n	0	8
2	m	1	0
3	q	0	0
4	r	6	0
5	p	4	3
6	s	0	7
7	t	0	0
8	o	5	0

Fig. 7.19.

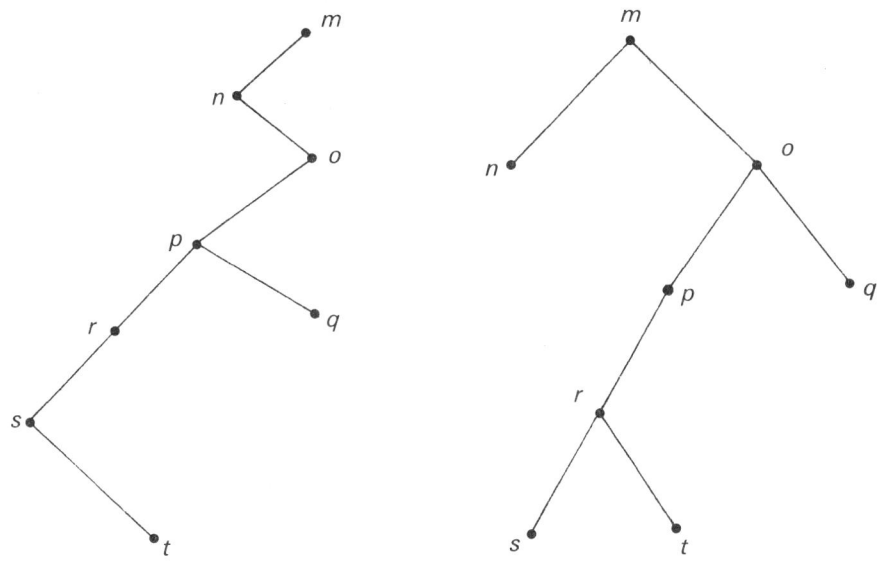

Fig. 7.20.

7.4 MINIMUM-WEIGHT SPANNING TREES

We want to consider a tree which can be thought of as a subgraph of a particular undirected connected graph. To see why we might wish to do this, think of a railway network, which might be modelled as a set of edges (routes) connecting vertices (towns); or a set of interconnected computers; or a network of telephone exchanges — any of these three might be thought of as a graph. What we would like to do is to reduce each of the three networks by taking out edges until the graph is *just* connected: that is, removal of a further edge would disconnect the graph (or make the corresponding railway/computer system/telephone system unworkable). This rather penny-pinching attitude to our original graph or system leaves us with a subgraph in the form of a *tree*, which still connects all vertices. Corresponding to the tree will be a workable — if inconvenient — system of connected towns, computers or telephone exchanges. We won't discuss the effectiveness of such pared-down systems here — clearly they could involve some very roundabout routes — but we will examine some of the ideas which arise in relation to such *spanning trees*, as they are called.

The first thing to ask is why this process should produce a *tree*, rather than some other form of connected graph. The answer is that we can only remove a certain type of edge without disconnecting the graph — namely an edge which is part of a *circuit*. If we were to remove an edge which was not part of a circuit, then we would disconnect the graph into two parts. This means that, by removing edges in order to break circuits, we will eventually obtain a connected graph with *no* circuits — that is, an unrooted tree.

Now that we have described what is meant by a spanning tree we will look at two related methods for deriving a spanning tree from a given connected graph. It turns out that algorithmically, *construction* (building up the tree) is easier than *trimming* (paring down the graph), though both methods involve selecting edges of the graph, whilst avoiding the creation of circuits. The basic reason for this is that it is easier to avoid edges which would create a circuit than to choose edges which will break a circuit; it's quite tricky to identify existing circuits, particularly since one edge might be part of several circuits.

To demonstrate the two methods, we'll construct a spanning tree from the graph shown in Fig. 7.21.

The first of our two methods — we won't call it an algorithm — is to select edges in turn from the graph and to construct the tree from these edges. The only restriction on the selection of edges is that a selected edge should not give rise to a circuit. The stopping rule is that we stop when the inclusion of a further edge is impossible without producing a circuit. One possible selection sequence is shown in Fig. 7.22. Note that disconnected sub-trees can appear whilst carrying out the process, as happens in this example.

The second method — shown in Fig. 7.23 — looks superficially similar. First select any vertex for the start, and consider it as the first vertex in the tree. Then choose (from the graph) an edge which is incident on this vertex, and put the edge (with its other vertex!) in the tree. This means that we now have two vertices in the tree. Continue by selecting edges one-at-a-time according to a similar rule: one vertex on the edge to be selected must be a vertex of the tree (clearly it is also a vertex of the graph), and the other vertex on the edge must be a member of the *graph* only. Stop when all vertices are contained in the tree. Notice that with this method we have a single connected tree at all stages, starting with the trivial case of a single vertex (and no edges), and finishing with all n vertices from the graph (and n-1 edges). It is worth pointing out that the simple rule 'one vertex is in the graph and the tree: the other vertex is in the graph only' is a formal statement which prevents the creation of loops.

Having established the idea of a spanning tree, we will now apply it to some situations involving edge-weighted undirected graphs. Such graphs arise from practical problems of the railway/computer/telephone network kind which we mentioned earlier. We can associate with each of the edges in such a network a numerical *weight*, which may represent a quantity such as distance, transmission cost, etc. Then an obvious question is whether we can find a set of edges linking all vertices in such a way that the total distance, total cost, etc. of the edges is a minimum. In other words, we want to find a *minimum-weight spanning tree*.

The methods for finding such a tree are *greedy algorithms* — that is, they involve making the 'greediest' or lowest weight selection at every stage in the construction of the tree. We have already made a passing reference to such greedy algorithms in Section 6.6, in connection with shortest and longest paths through a directed graph. In that context they were not particularly helpful, but, as we will now see, they can be used very successfully to modify the two methods described above for finding a spanning tree, and yield two important algorithms for the construction of minimum-weight spanning trees.

Modification of the two methods for finding a spanning tree is very simple: if at any stage of the process there is more than one edge of the graph which *could* be selected for the tree, choose the edge which has *minimum* weight. This procedure

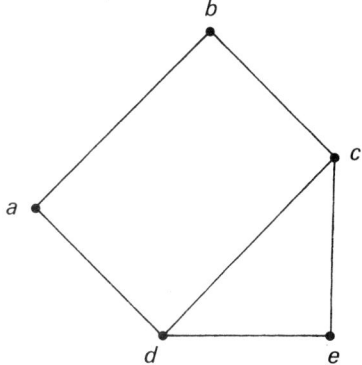

Fig. 7.21.

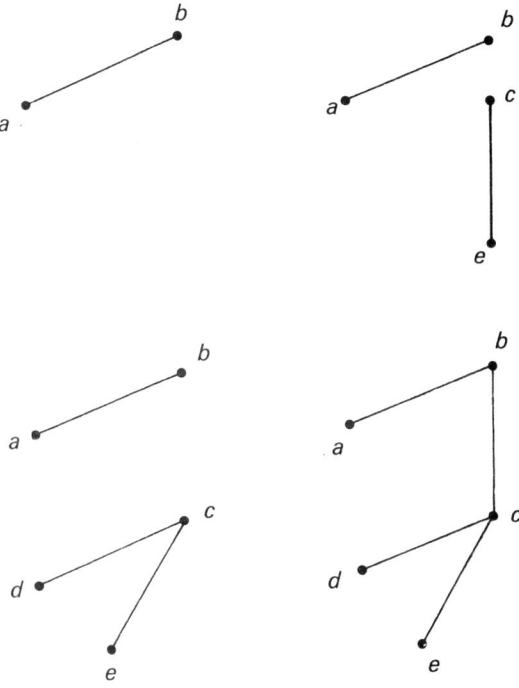

Fig. 7.22.

will guarantee a minimum-weight spanning tree. (Whether or not you think that the result is obvious, it is in fact quite difficult to prove.) The first of the algorithms, corresponding to the case where we took edges one by one, subject to *not* forming a

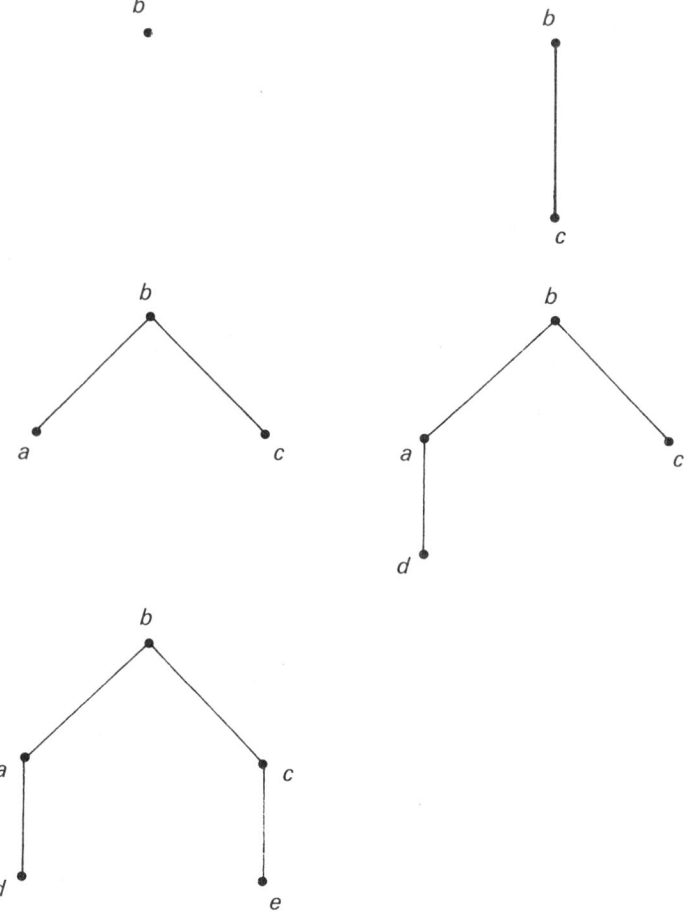

Fig. 7.23.

circuit, is known as Kruskal's algorithm: the second, where we maintained a (connected) tree throughout, is known as Prim's algorithm. We will first show how the two algorithms are applied to the weighted graph shown in Fig. 7.24 — we'll do this fairly informally, by eye as it were. Then we will take Prim's algorithm and show how we can formalize it to make it suitable for automatic processing. Finally we'll give a proof that one of the algorithms — the one due to Kruskal — does produce a minimum-weight spanning tree.

If we apply Kruskal's algorithm to the graph, we get a sequence as shown in Fig. 7.25, which can be described as follows:

(1) The edge of lowest weight is $\{a,c\}$, with a weight of 1; so this edge becomes part of the tree.
(2) The edge $\{a,d\}$, is the next (weight=2) so this edge is added.
(3) The next lightest edge is $\{c,d\}$, (weight=4), but we can't attach this edge to the

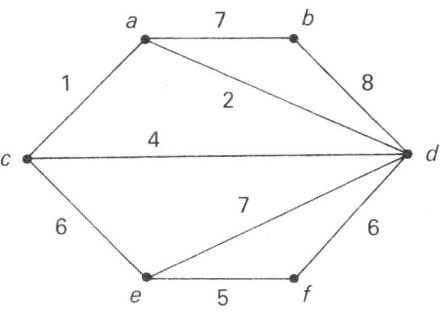

Fig. 7.24.

tree as it would form a circuit $\langle a,c,d,a \rangle$; so we choose the next lightest edge $\{e,f\}$ (weight=5). (Notice that at this point in the construction there are two separate sub-trees.)

(4) Now we can choose either of the edges $\{c,e\}$ or $\{d,f\}$ (both of weight 6): we will arbitrarily take $\{d,f\}$.
(5) We can't choose $\{c,e\}$ next, as it would complete a circuit; so we take the next lightest edges $\{a,b\}$ and $\{d,e\}$ (both of weight 7). Since $\{d,e\}$ would form a circuit, we must choose $\{a,b\}$. This completes a spanning tree of weight 21. Note that there are at least two possible criteria for deciding that the process has terminated: one is that all vertices are now included in the tree, and another is that the tree consists of five edges, that is, one fewer than the number of vertices. Can you think of a third reason why the process can't continue further?

The application of Prim's algorithm to the graph is shown in Fig. 7.26, and described below.

(1) We start arbitrarily, without reference to edge weights, from the vertex e, which becomes part of the tree.
(2) If we look from e along edges to adjacent vertices, we see that the lightest edge is $\{e,f\}$ with weight 5. This is the greediest choice possible at this stage, so the edge $\{e,f\}$ — and hence vertex f — are added to the tree.
(3) Looking from vertices e and f we see that $\{e,c\}$ and $\{f,d\}$ are equally 'lightest'; we will take $\{e,c\}$, bringing in vertex c.
(4) Now we can look along edges from vertices c, e and f. Since $\{c,a\}$ (with weight 1) is the lightest, we put this edge, and vertex a, into the tree.
(5) Now looking from vertices a, c, e and f we see that $\{a,d\}$ (weight=2) is the next choice.
(6) Finally, looking from vertices a, c, d, e and f we find that $\{a,b\}$ (weight=7) is the best choice, so vertex b is now connected as well. The process terminates since there are no vertices which are part of the graph whilst not being part of the tree; in other words, there are no vertices to look for.

Notice that our application of this algorithm has led to a different spanning tree, though one with the same minimum weight of 21.

Obviously it is easy to make mistakes in constructing spanning trees — we rely on

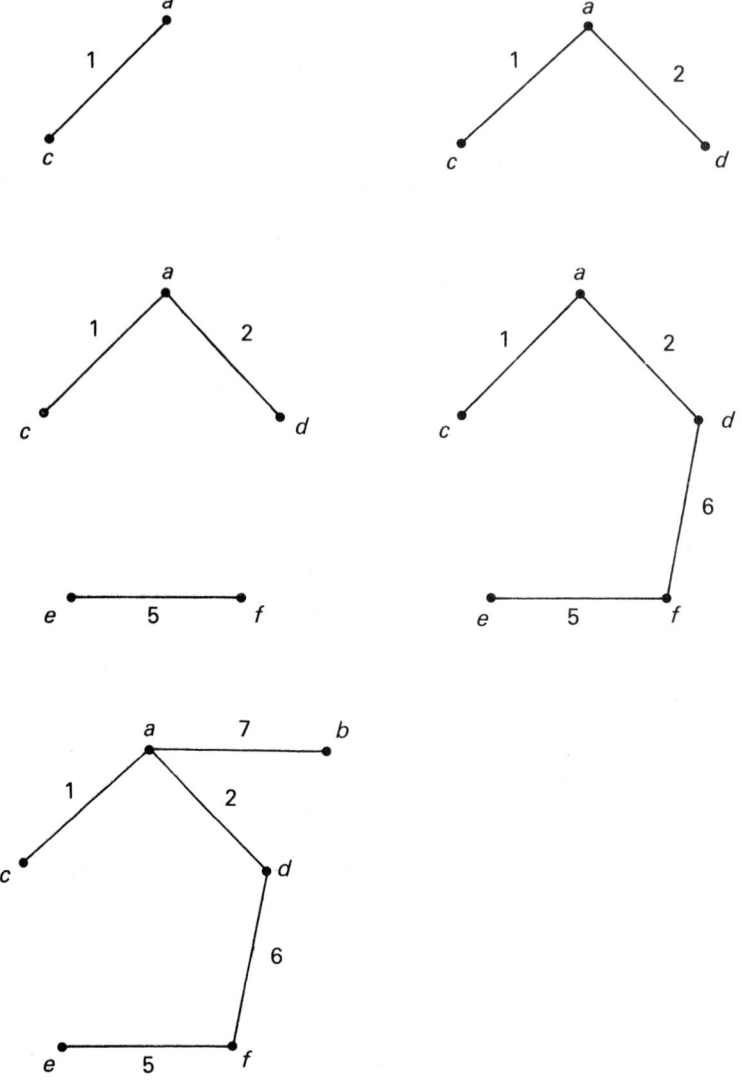

Fig. 7.25.

our powers of observation, and in a large graph it can become difficult to spot that a long circuit is about to be completed, and very easy to miss a current lightest edge. However, the form of Prim's algorithm makes it possible to carry out the process systematically, using a matrix representation of the graph. This is shown in Table 7.1: for simplicity we include each stage as a separate entry, though in practice we could work with just the first entry, making amendments to the picture as necessary. We will repeat the construction of the tree we have just produced. The same sequence of pictures applies as before (Fig. 7.26), but this time we will refer to the table.

(1) Choose vertex e, as before.

Sec. 7.4] Minimum-weight spanning trees 175

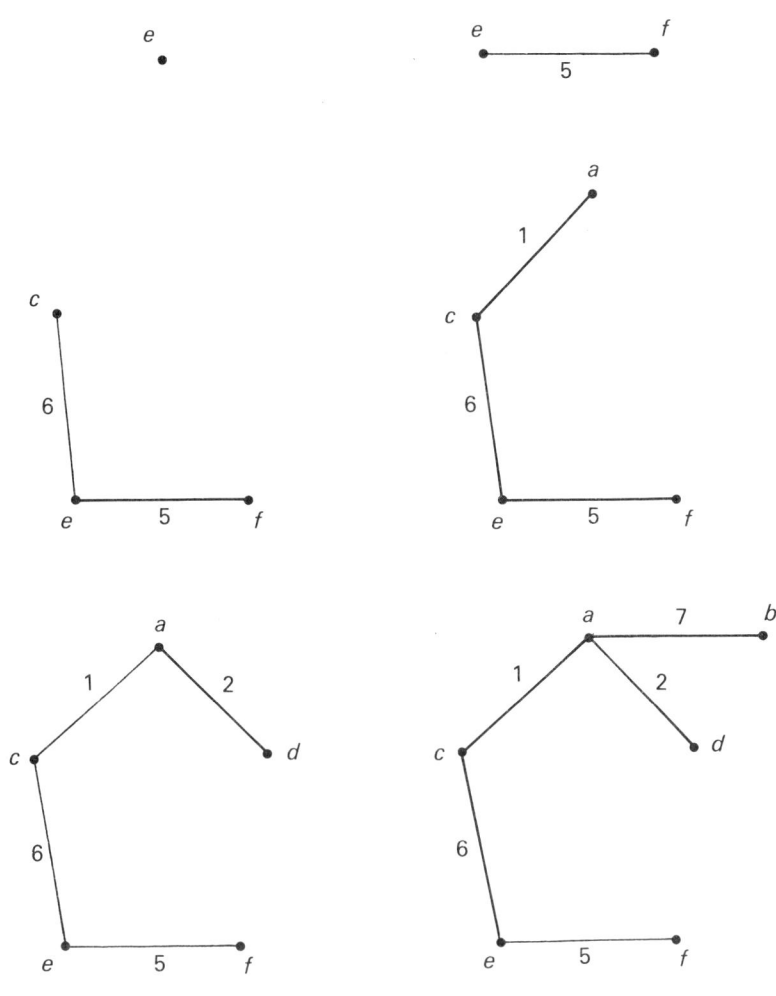

Fig. 7.26.

(2) We *mark* the column corresponding to e, to indicate that we may look from vertex e in the graph along edges of the graph. At the same time *delete* entries in the row corresponding to e, since at no time is it permitted to look along an edge of the graph *towards* vertex e — we don't want to allow any cycles to be formed. Looking at entries in the marked column, we see that the lowest value is 5, which corresponds to edge $\{e,f\}$. This means that edge $\{e,f\}$ is added to the tree.

(3) We mark the column corresponding to vertex f, and delete the row entries corresponding to f. We can now look down the e and f columns to find the lightest edge: there are two cases of a minimum value of 6, and we will take the one corresponding to the vertex c (and the edge $\{c,e\}$).

(4) We mark column c and delete the entries in row c; and look for the smallest entry in the three marked columns c, e and f. The value 1 is the smallest, and so edge $\{a,c\}$ is accordingly chosen.

176 Trees [Ch. 7

Table 7.1

(1)

	a	b	c	d	e	f
a	—	7	1	2	—	—
b	7	—	—	8	—	—
c	1	—	—	4	6	—
d	2	8	4	—	7	6
e	—	—	6	7	—	5
f	—	—	—	6	5	—

(2)   *

	a	b	c	d	e	f
a	—	7	1	2	—	—
b	7	—	—	8	—	—
c	1	—	—	4	6	—
d	2	8	4	—	7	8
e	—	—	—	—	—	—
f	—	—	—	6	5	—

                                             *

(3)   *   *

	a	b	c	d	e	f
a	—	7	1	2	—	—
b	7	—	—	8	—	—
c	1	—	—	4	6	—
d	2	8	4	—	7	6
e	—	—	—	—	—	—
f	—	—	—	—	—	—

                                  *   *

(4)   *   *   *

	a	b	c	d	e	f
a	—	7	1	2	—	—
b	7	—	—	8	—	—
c	—	—	—	—	—	—
d	2	8	4	—	7	8
e	—	—	—	—	—	—
f	—	—	—	—	—	—

                           *   *   *

(5)   *   *   *   *

	a	b	c	d	e	f
a	—	—	—	—	—	—
b	7	—	—	8	—	—
c	—	—	—	—	—	—
d	2	8	4	—	7	6
e	—	—	—	—	—	—
f	—	—	—	—	—	—

     *   *   *   *

(6)   *   *   *   *   *

	a	b	c	d	e	f
a	—	—	—	—	—	—
b	7	—	—	8	—	—
c	—	—	—	—	—	—
d	—	—	—	—	—	—
e	—	—	—	—	—	—
f	—	—	—	—	—	—

    *   *   *   *   *

(5) We mark column a and delete entries in row a. The smallest value in the marked columns a, c, e and f is the 2, corresponding to the edge $\{a,d\}$, which accordingly comes into the tree.

(6) Column d is now marked and the entries in row d are deleted. Looking down the

columns a, c, d, e and f we find the smallest entry is 7, and corresponds to edge $\{a,b\}$, so vertex b now gets attached to the tree. We see that marking column b and deleting the row entries for b will terminate the process — since all columns are now marked, and all entries in the table have been deleted. This gives us a simple stopping rule, as well as a checking rule: when all the columns have been marked, there should be no entries left in the table.

You might like to find a reasonable justification for the apparent redundancy of information in the original table. The matrix was symmetric, with a leading diagonal of 'empties': why can't we modify the algorithm to work on the upper right triangle of the matrix, since it contains all the data we need?

A proof of the correctness of Kruskall's algorithm

This may look like an unnecessary complication, but it is offered for two reasons: first as an example of proof by contradiction, rather more taxing than the examples you have encountered earlier in the book, and secondly as a part of the validation of an algorithm — currently a luxury among programmers, but certain in the future to become a necessary part of systems practice. See whether you can understand the argument set out here, even if you feel you could not reproduce it.

Remember that the effect of Kruskall's algorithm is to produce a tree (why must it be a *tree*?) with n-1 edges from a connected graph with n vertices. If we write out the list of edges in ascending order of weight (exactly the order in which Kruskall's algorithm selects them), we can represent the Kruskall tree as $K = \langle e_1, e_2, e_3 \ldots, e_{n-1} \rangle$, where the weight of each element (edge) will be less than (or equal to) the weight of its right-hand neighbour, so that weight $(e_1) \leq$ weight $(e_2) \leq \ldots$.

We want to show that the tree produced by Kruskall's algorithm from a given connected graph of n vertices is a minimum-weight spanning tree. We will call the tree K, and for the purposes of this proof, which is a proof by contradiction, we will start with the assumption that the Kruskall tree K is *not* a minimum-weight spanning tree. Instead we are going to suppose the existence of a set of minimum-weight spanning trees with a smaller weight than K; if there is more than one such minimum-weight tree we will choose the tree B, which has the largest number of edges in common with K.

In brief, K and B are spanning trees, and we suppose B to be of minimum weight. We can write the fact that the weight of B is less than or equal to the weight of K as

$w(B) \leq w(K)$.

Now for a start, there must be at least one edge in K which is not in B, otherwise the two trees would be the same tree. What we are going to do (in our imaginations at least!) is to look through the edges in the list K from left to right — that is, in ascending order of weight — until we meet an edge e_K, in the tree K which is *not* an edge in the tree B. You might like to sketch out a small connected graph and mark in two different spanning trees to help you visualize it — don't bother putting any weights in though. Next we attach this edge e_K to the list of edges which makes up the tree B (you might like to sketch this as well). Now since B was defined as a spanning tree, an obvious consequence of adding this extra edge to the tree B is to form a *cycle* (of some length between 3 and n); so that B with its extra edge is no longer a tree. To get our augmented B back to a tree form we must remove an edge from the cycle we've just created. The cycle must contain at least one edge which is *not* in the Kruskal tree K, since K does not contain a circuit. Call this edge e_B and remove it from the augmented B. Now let's consider how this process of attaching e_K and removing e_B changes the weight of B, our supposed minimum-weight tree. For convenience denote by B' the version of B with e_K added and e_B deleted. Clearly

$$w(B')=w(B)+w(e_K)-w(e_B).$$

We can look at this equation in two ways. Firstly we claimed that B was a minimum-weight spanning tree, so

$$w(B) \leq w(B')$$

which means in turn that

$$w(e_K) \geq w(e_B).$$

Secondly we recall the Kruskal rule for selecting edges for the spanning tree K: select the edge of minimum weight in the graph, provided the selected edge does not complete a cycle. Recall further that edges e_1 up to e_{K-1} were all members of both K and B. However, e_K must have a smaller weight than e_B otherwise the Kruskall algorithm would have selected e_B for the tree K, when in fact e_K was chosen. For this reason we can state that

$$w(e_B) \geq w(e_K)$$

From this we can see that $w(e_B)=w(e_K)$, so that $w(B)=w(B')$. So B' has the same weight as B, but has one more edge — e_K — in common with K. This is the contradiction we were looking for: our original assumption was that B was the minimum-weight tree which differed from K in the smallest number of edges. We have just produced a tree, B' with minimum weight, which differs from K by one less edge than B. We can therefore say that our initial assumption must be wrong, and so there does not exist a spanning tree of lower weight than K.

Exercise 7.4.1
Consider the following modification of the algorithm. Start with the original graph, and remove the *heaviest* edge first, then the next heaviest, and so on — always subject to keeping the graph connected. The stopping rule is that all circuits have been removed. Does this version of the algorithm work?

7.5 BINARY POSITIONAL TREES REVISITED

For the rest of this chapter we are going to look at some useful properties of the binary positional tree, which have interesting applications. We will see how a binary

Sec. 7.5] Binary positional trees revisited

positional tree can be applied to searching and sorting, and to operations in arithmetic and algebra. As with our work on graphs (Chapters 5 and 6) and with earlier work in this chapter, we can take a diagrammatic approach, whilst remembering that this could be formalized, if we wished. To follow this section you should be familiar with the representation of a binary positional tree as a *doubly linked list* described in Section 7.3.

In this section all the ideas relate to the notion of a *tree search*. A tree search consists of working systematically through the tree by traversing its edges and the associated vertices. The purpose of this search is to gain access to a *particular* (predetermined) vertex, or sometimes to *all* vertices. By the phrase 'to gain access' we mean to obtain — and process in some way — whatever information is contained by the vertex (this information could be as complex as a complete student record, or an index to a set of computer files, but in our examples will be something much simpler — like a number or a letter, or simply the vertex label itself). Our processing will also be very simple: by the verb 'process' we will mean the operation of printing (or outputting in some form) the vertex label.

We will find that there are a number of closely related systematic ways of making a complete search (or *traverse*) of the tree, and that — not surprisingly — different search methods produce different sequences or lists of output. When we have looked at three particular search methods, we will examine some applications.

First of all we will give a simple topological (or 'geographical') rule for navigating the tree in its pictorial form. After that we'll decide on a rule for processing (printing out in our case) the vertices of the tree, but initially we need to be certain our rule will guarantee that all vertices are visited at least once.

The rule is as follows. Start from a position above the root vertex, and walk round the tree, close to the edges, *so that the tree is always on your immediate left*; in effect, take an anti-clockwise direction. Whenever you pass a vertex on your left, this vertex is *visited* (don't forget that your left is relative to the direction you are facing at any given time!). The walk terminates when the root vertex is visited for the *third* time, since at this point you will have done a complete circuit round the tree — to continue would be to repeat the journey! The course of such a journey is shown in Fig. 7.27, where you will see that all vertices received at least one visit. In fact vertices a, b, j, and k all received three visits; vertices c and i received two; and g, f, d, h, and e received just one. The reason for the different numbers of visits is obvious from looking at the tree, but the best way to think of it is to ask what causes return visits to a vertex. The answer is that after a *first* visit to a vertex, one of two cases is possible: either the vertex has no children, or the vertex has children. If the vertex has no children then it can't receive another visit. If the vertex has one or two children, each child must at some point be visited, and after each visit there must be a revisiting of the parent; so the number of visits to a given vertex is made up of one visit for the vertex itself, and one return visit for each of its children. The visiting sequence in Fig. 7.27 is $\langle a,b,k,g,k,f,k,b,d,b,a,j,c,i,h,i,c,j,e,j,a \rangle$, and you can check from this that a vertex with m children is visited $m+1$ times.

This is a perfectly satisfactory way of gaining access to all vertices, but if we think of the tree in a slightly different way we'll find that our navigation method will produce a much more useful visiting sequence; in fact we can ensure that *all* vertices receive *three* visits. Recall the representation of a binary positional tree in linked-list

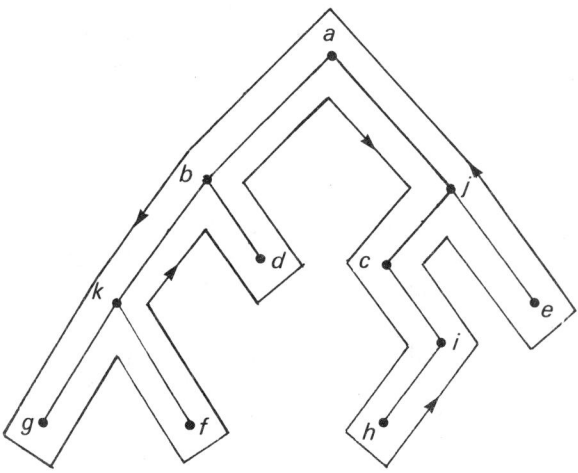

Fig. 7.27.

form. We can repeat the explanation of the last paragraph, but this time suppose that we think of it in terms of linked lists, where a vertex consists of a left pointer, the record, and a right pointer. We could say that any vertex represented in this way has a left child and a right child, as indicated by the two pointers. The three visits are the first visit after which the walk follows the *left* pointer, but must eventually return to the vertex for a second visit. The walk then follows the *right* pointer, but must eventually return for a third (and final) visit to the vertex. This is plainly the case for vertices a, b, j and k, but it's also true for the other vertices, if we consider them to have left and right pointers as well. The left pointer from c, the right pointer from i, and *both* pointers from each of d, e, f, g and h, all have value 0; nevertheless any such pointer has to be examined, before it can be known that it *is* a zero. We can think of this examination as a (short) excursion from the vertex along the pointer, followed by a return visit to the vertex.

We can include these zero pointers on the tree diagram simply by putting in dotted lines to represent them (Fig. 7.28). As they are pointing to non-existent vertices, the pointers are not strictly edges (since an edge is always defined by *two* vertices). However, their inclusion brings the picture representation closer to our linked-list representation. Check that the sequence we get from this picture is

$$\langle a,b,k,g,g,g,k,f,f,f,k,b,d,d,d,d,b,a,j,c,c,i,h,h,h,i,i,c,j,e,e,e,j,a \rangle.$$

You can see that every vertex receives three visits.

The fact that we can access/process all of the vertices of a tree is not in itself remarkable; we can usually unload the contents of a data structure — whether it be a book of raffle tickets or the payroll file for Ford Motors. In order to capitalize on the specific properties of our *tree*, we will look at three *systematic* ways of processing the vertices of the tree (there are others, but these three have particular relevance to computing and information technology in general). Our method of traversing the

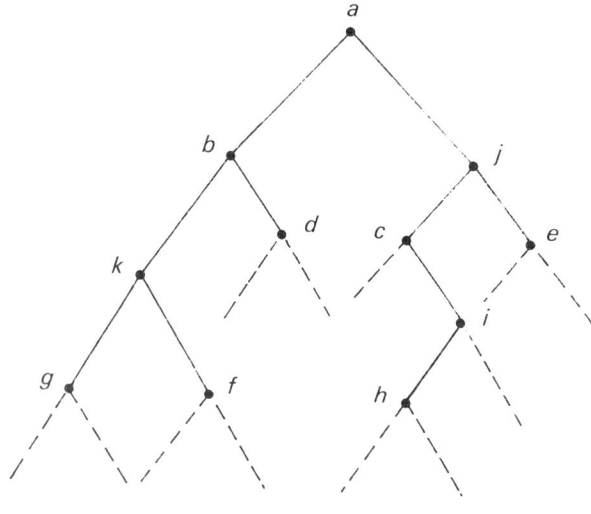

Fig. 7.28.

tree gave us *three* visits to each vertex, and hence we have three different opportunities to process the information contained by each vertex. Simply stated — and as you can probably guess — we are going to examine the three cases where we process respectively on the first visit; on the second visit; and on the third visit.

The full sequence of visits was

$$\langle a,b,k,g,g,g,k,f,f,f,k,b,d,d,d,b,a,j,c,c,i,h,h,h,i,i,c,j,e,e,e,j,a \rangle.$$

If we process each vertex on the *first* visit we get a sequence

$$\langle a,b,k,g,f,d,j,c,i,h,e \rangle.$$

This is often called a *pre-order* search, for obvious reasons — and it's worth looking again at Fig. 7.27, where you will see that the effect is to go as far down the tree as possible, before reluctantly coming back up for air! Another name for the search is 'depth-first search'.

If we process on the *second* visit — what is often called an *in-order* search — we get another sequence:

$$\langle g,k,f,b,d,a,c,h,i,j,e \rangle.$$

The sequence obtained when we process on the third and last visit is

$$\langle g,f,k,d,b,h,i,c,e,j,a \rangle$$

and this is often called a *post-order* search.

Example
Write out the pre-, in-, and post-order search sequences of the binary positional tree shown in Fig. 7.29.

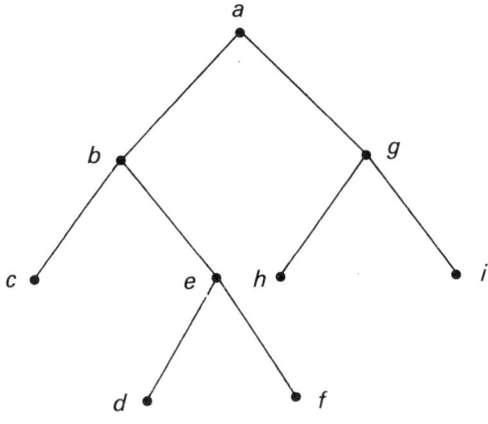

Fig. 7.29.

Answer
The pre-order sequence is $\langle a,b,c,e,d,f,g,h,i \rangle$
the in-order sequence is $\langle c,b,d,e,f,a,h,g,i \rangle$
and the post-order is $\langle c,d,f,e,b,h,i,g,a \rangle$,

We are going to look at a simple application of the three methods of tree-searching, which relates each method to the evaluation of an arithmetical or algebraic expression.

An example of an algebraic expression is $b \uparrow 2\text{-}4ac$ which you may have met when dealing with quadratic equations. When the expression is written in this particular way, it is assumed that you know where the missing multiplication signs should go. You also need to know that the exponentiation operation $\uparrow$ takes precedence over the next operation of subtraction $-$, and that the invisible multiplication operations between the 4, the a and the c have precedence over the $-$, though not over the $\uparrow$. To avoid any ambiguity we could explicitly put in all the operators (no invisible multiplications!) and also *parenthesise* the expression (no arguments about precedence). This would give the following *fully parenthesized expression*

$$((b \uparrow 2) - (4*(a*c)))$$

Then the rules for evaluation are unambiguous: if you have two operands — a and c for example — in a bracket and separated by an operator, then evaluate the result of the operation and remove the brackets.

The fully parenthesized expression lends itself rather well to structuring as a binary positional tree, for the following reason. Each pair of brackets contains three parts, an *algebraic expression* followed by an *operator* followed by a second *algebraic expression*; this structure can be reflected in the tree by placing the operator as parent, and the algebraic expressions as its two children.

If we carry out this process with the expression above, working from the outermost brackets, we see that we have the algebraic expression $(b \uparrow 2)$ followed by the minus operator followed by the algebraic expression $(4*(a*c))$. So we begin by labelling the root of the tree with the minus operator, the left child with $(b \uparrow 2)$ and the right child with $(4*(a*c))$.

Now we can dismantle the left child in the same way, showing $\uparrow$ as parent, b as left child and 2 as right child. This is as far as we can go; the algebraic expressions have become single elements, or *operands*. Returning to the right child of the minus operator, we decompose it into a parent vertex * with respective left and right children 4 and $(a*c)$. Finally $(a*c)$ can be decomposed into a parent * with children a and c. The complete tree representation is shown in Fig. 7.30.

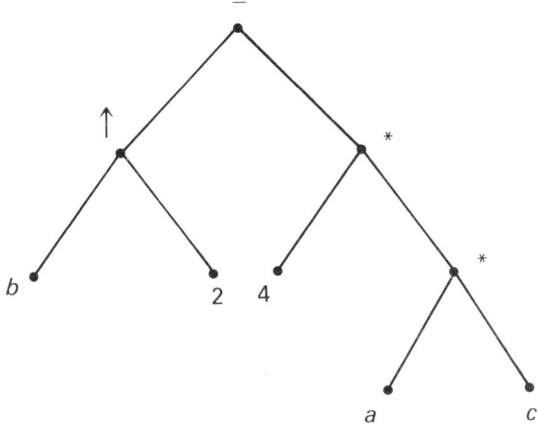

Fig. 7.30.

Now let's look at a numerical version — an arithmetic rather than an algebraic expression. Suppose that in the expression above $a=2$, $b=8$ and $c=5$. The expression becomes $8 \uparrow 2 - 4*2*5$ (we can't omit the * signs any longer!), and the fully parenthesized version is $((8 \uparrow 2) - (4*(2*5)))$ or alternatively $((8 \uparrow 2) - ((4*2)*5))$. The first of these two is shown as the first tree in Fig. 7.31.

This expression can be evaluated from the bottom up, by replacing each operand–operator–operand sub-tree by its actual *value*. The effect of this replacement is to produce a new operand with the given value. Also three vertices of the tree are replaced by one vertex. As you can see, this means that we can work upwards from the leaves of the tree, replacing any arithmetical operation with its result (see Fig. 7.31).

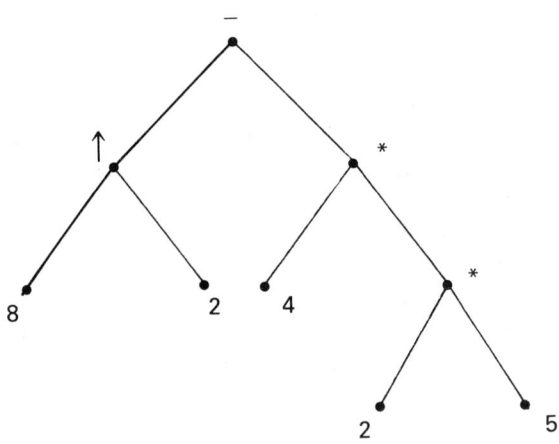

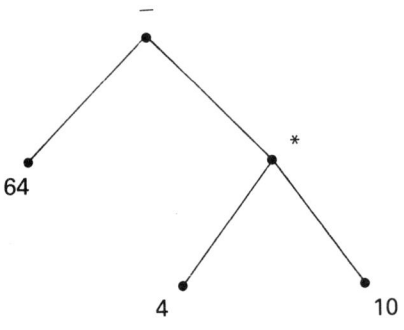

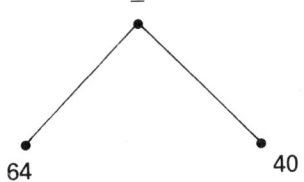

Fig. 7.31.

Sec. 7.5] **Binary positional trees revisited** 185

In this way the sub-tree 8 ↑ 2 is replaced by its value of 64, and the sub-tree 2∗5 is replaced by the value 10. Next we can replace the sub-tree 4∗10 which has value 40. This leaves us with just one sub-tree 64–40, giving a value of 24, which is the value of original expression.

Example
Draw a tree to represent the fully parenthesized expression (((8+10)/6)∗(12−(2 ↑ 3))) and evaluate it.

Answer
The tree is shown in Fig. 7.32. The value of the expression is 12.

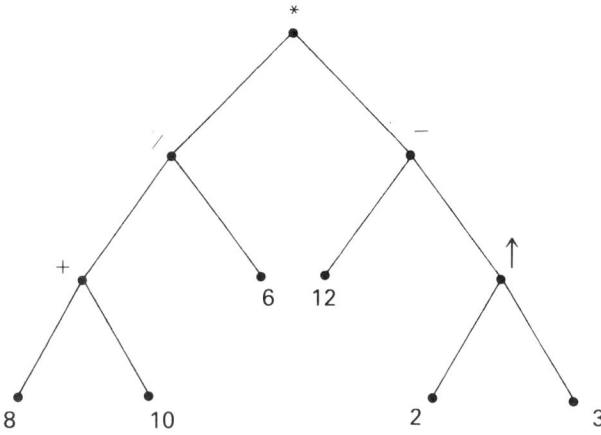

Fig. 7.32.

While we have been carrying out these algebraic and arithmetical evaluations we have been using what you might consider to be 'ordinary arithmetic' — that is, expressions of the form operator–operand–operator, which in turn produces another operand. This form of arithmetic is known as the *infix* form. Infix form has been fairly standard throughout history, but is certainly not the only form; if we apply our three types of tree search to the tree of Fig. 7.31 we will find that two new forms of arithmetic appear.

The pre-order search — that is, the search in which we process on the first visit — gives

− ↑ 8 2∗4∗2 5

and this expression can be evaluated if we make a change in the order of our arithmetic. If we decide to replace the familiar operand–*operator*–operand order of calculation with the alternative order of *operator*–operand–operand, then we get a new arithmetic called *prefix* arithmetic. So the commonplace 2+2, which is an infix expression, has the prefix representation +2 2 (the value of the expression is still 4, of course); a division such as 21/3 becomes /21 3 with a value of 7, and so forth.

The evaluation of a prefix expression can be described by the following action diagram:

```
┌Repeat
│   Read the expression left to right.
│   ┌If you meet a sequence of three characters in the form operator–operand–o-
│   │perand evaluate the sequence and replace with its value
│   └Endif
└Until the sequence has just one element
```

Let's apply this to the example above. Reading from the left, we meet the sequence ↑ 8 2, which we read as 'the exponentiation of 8 by 2', and which has a value of 64; so now the expression is −64∗4∗2 5. We continue reading left to right, and find the sequence ∗2 5, which gives the value 10: so the expression becomes −64∗4 10. As we have reached the right-hand end of the expression, and the expression is not yet a single element, we start again from the left. This time we can evaluate ∗4 10 as 40; the expression becomes −64 40; and again we are at the right-hand end. If we start again from the left, we have just one evaluation to carry out, giving us the single element, 24, which is the value of the expression.

Now let us look at the effect of the two other search methods on the same tree. But, first, try this example to make sure you understand the prefix idea.

Example
Write out the result of a pre-order search on the tree shown in Fig. 7.32 and evaluate the expression.

Answer
The result of the pre-order search is ∗/+8 10 6−12 ↑ 2 3
and its value is obtained in the sequence ∗/18 6−12 ↑ 2 3
∗3−12 ↑ 2 3
∗3−12 8
∗3 4
12

Note that we have carried out just one operation on each line, though the evaluations + 8 10 and ↑ 2 3 could have been (and according to the action diagram, should have been) carried out on the same line.

The *in-order* search of the tree is the case of second-visit processing (which is quite easy to write down directly from the tree, as long as you remember that in effect, each leaf of the tree gets three visits). The result of the search on the tree in Fig. 7.31 this time is

8 ↑ 2−4∗2∗5

which turns out to be something of a disappointment. The expression is the original infix expression, but *without* the brackets which would enable us to evaluate it with certainty. We could get the same (unparenthesized) expression from a number of different fully parenthesized versions, for example $(((8 \uparrow (2-4))*2)*5)$, which has value 0.15625.

The *post-order* search of the tree (that is, third-visit processing) gives the following expression:

$$8\ 2 \uparrow 4\ 2\ 5 * * -$$

The evaluation of this expression is similar to the evaluation of the pre-order expression, with a single modification. We amend the order of our arithmetic to the form operand–operand–*operator*: that is, our arithmetic becomes what is called *postfix* arithmetic. The old infix expression 2+2 becomes the postfix expression 2 2+, just as 2⁄3 becomes 2 3⁄, and so on.

The action diagram for evaluation of the postfix expression differs from that for the evaluation of the prefix expression by the re-ordering of one line (the re-ordered bit is italic):

⌐Repeat
⎢　　Read the expression left to right
⎢　　⌐If you meet a sequence of three in the form *operand–operand–operator*
⎢　　⎢ *evaluate the sequence and replace with its value*
⎢　　⌊Endif
⌊Until the sequence has just one element

The expression $8\ 2 \uparrow 4\ 2\ 5 * * -$ is evaluated according to this algorithm as follows. The sequence $8\ 2 \uparrow$ is evaluated as 64, and the expression becomes $64\ 4\ 2\ 5 * * -$. The next sequence of three we meet is 2 5∗, which we can evaluate as 10. The expression is now 64 4 10∗ −, and we reach the right-hand end without being able to evaluate any more operator–operator–operand sequences.

So we have to go back to the left of the expression, and the first sequence we can evaluate is 4 10∗, with value 40. This leaves the expression as 64 40−. As we have reached the right-hand end again, we return to the left and evaluate 64 40− as 24, which is the final value of the expression.

Example
Show that a post-order search on the tree shown in Fig. 7.32 gives rise to the expression 8 10+6/12 2 3 ↑ −∗ and verify that the value of the expression is again 12.

7.6 TREE-LESS EVALUATIONS

This may seem an eccentric choice of topic in a chapter about trees, but if you've understood the last section, we hope that your new knowledge of trees will give you an insight into a method that doesn't use them.

Arithmetic expressions written in pre-order and post order form are sometimes given the respective names of *Polish* and *reverse Polish* expressions, after the nationality of their inventor, Jan Lukasiewicz. As some calculators — and many compilers — make use of reverse Polish, we'll show how to convert a straightforward arithmetic or algebraic expression into reverse Polish.

We must take into account the fact that an arithmetic or algebraic expression normally contains just enough brackets to avoid ambiguity — so we must allow for the rules of precedence of operators. The descending order of precedence is $\uparrow$, $*$, $/$, $+$, $-$, ( ). The low priority of an opening bracket will be understood if you remember the rule drilled into you at school 'Work out the things *in* the bracket first!' This means that the actual bracket is implemented last of all.

Next we need to set out a space in which to carry out the conversion. We have three parts to the space; first the *input string* which is in effect our initial arithmetic expression, with a start and finish marker (looking like inverted drawing pins $\bot$). We have an *output string,* which is empty initially, but in which the reverse Polish expression will be built up. Finally we have a temporary *stack* onto which various operators will be placed, when removed from the input string. As an example, the top of Table 7.2 shows the initial arrangement for evaluating the expression

$$\bot 7+2*(x+5)\uparrow 3\bot$$

(note the algebraic operand x; we can apply our Polish and reverse Polish to infix arithmetic *and* to infix algebra — and indeed to Boolean expressions, if we wish).

There are six rules for carrying out the conversion to reverse Polish, only one of which is longer than a sentence. When reading them, you should remember that only the leftmost element of the input string, and only the top element of the stack, is 'visible' at any time in the process.

(0) Put the initial 'drawing pin' on the stack.
(1) An *operand* on the input string is immediately placed on the output string.
(2) (a) If an *operator* on the input string has a *higher* priority than the operator on the *stack* then stack the operator from the input string; otherwise
   (b) pop the operator on the stack to the output string.
(3) Always put an opening bracket on the stack.
(4) If there is an opening bracket on the stack, and a closing bracket on the input string, then delete both brackets.
(5) If there is a drawing pin on the stack, and a drawing pin on the input string, then the output string should now be a valid reverse Polish expression.

The complete coversion is shown in Table 7.2: notice that the *input string* is on the right, (so that we read it from left to right!): the stack is in the middle; and the output string is on the left.

Example
Produce the reverse Polish equivalent to the infix algebraic expression $(x+y)\uparrow n-2*z$ using the procedure outlined above.

Answer
The expression is $xy+n\uparrow 2z*-$

Table 7.2

Output string	Stack	Comment	Input string
			$\perp 7 + 2 * (x + 5) \uparrow 3 \perp$
	$\perp$		$7 + 2 * (x + 5) \uparrow 3 \perp$
		rule 0	
7			$+ 2 * (x + 5) \uparrow 3 \perp$
	$\perp$	rule 1	
7	$+$ $\perp$		$2 * (x + 5) \uparrow 3 \perp$
		rule 2(a)	
7 2	$+$ $\perp$		$* (x + 5) \uparrow 3 \perp$
		rule 1	
7 2	$*$ $+$ $\perp$		$(x + 5) \uparrow 3 \perp$
		rule 2(a)	
7 2	$($ $*$ $+$ $\perp$		$x + 5) \uparrow 3 \perp$
		rule 3	
7 2 x	$($ $*$ $+$ $\perp$		$+ 5) \uparrow 3 \perp$
		rule 1	
7 2 x	$+$ $($ $*$ $+$ $\perp$		$5) \uparrow 3 \perp$
		rule 2(a)	
7 2 x 5	$+$ $($ $*$ $+$ $\perp$		$) \uparrow 3 \perp$
		rule 1	
7 2 x 5 $+$	$($ $*$ $+$ $\perp$		$) \uparrow 3 \perp$
		rule 2(b)	
7 2 x 5 $+$	$*$ $+$ $\perp$		$\uparrow 3 \perp$
		rule 4	
7 2 x 5 $+$	$\uparrow$ $*$ $+$ $\perp$		$3 \perp$
		rule 2(a)	
7 2 x 5 $+$ 3	$\uparrow$ $*$ $+$ $\perp$		$\perp$
		rule 1	
7 2 x 5 $+$ 3 $\uparrow$	$*$ $+$ $\perp$		$\perp$
		rule 2(b)	
7 2 x 5 $+$ 3 $\uparrow$ $*$	$+$ $\perp$		$\perp$
		rule 2(b)	
7 2 x 5 $+$ 3 $\uparrow$ $*+$	$\perp$		$\perp$
		rule 2(b)	

7.7 SORTING BY TREE

Now we'll look at a simple but ingenious use of binary positional trees in the procedure of sorting, which is very common in data processing. This will give us a chance to use an in-order (or second-visit) search. In outline: we take a string of numbers, which as far as we know is in no particular numerical order, and we construct a tree with the numbers as *vertices,* according to a simple construction rule. The *deconstruction* rule — which we use for getting an ordered list from our tree — turns out to be a straightforward in-order search.

Suppose we start with the list $\langle 17,14,22,15,5,37,12,20 \rangle$ as an example. We are going to take the numbers in turn from the head of the list, and decide where each one should be placed in the tree. If you like, you can think of the procedure as a method of *growing* a tree starting with the root. So the first element in our list, 17, becomes the *root* of the tree. Every subsequent member of the list can be thought of as being processed *down* the tree (from the root) until it can be processed no further, at which point it becomes a *leaf* of the tree.

Once we have established the root of the tree, the rule for processing further elements onto the tree is very simple. We read each element in turn from the list, and compare it first with the root vertex of the tree. If it is *less* than the root element, it is directed to the *left* child vertex of the root; if *greater*, it goes to the *right* child vertex. Then the comparison is repeated, and in this way the element is passed down the tree until it is being compared with a leaf. After this, no further comparisons can be made, and the element itself becomes a leaf of the tree.

This will become clearer if we process the next few elements in the list above. We have already made 17 the root of the tree. We now read the next element from the list, 14, and compare it with the value of the root. As 14 is less than 17, it becomes the left child vertex of 17. Next we take the 22, which is greater than 17, and so is established as its right child vertex.

The next element from the list is 15, and this will need more than one comparison before finding its place as a vertex. On the first comparison with the root vertex 17, it will be sent left. It then meets the 14, comparison with which sends it right; there being no further vertices for comparison, the 15 is then established as the right child of the 14.

You should now look at Fig. 7.33, and make sure that you can follow the trail of the remaining elements 5, 37, 12, 20, as they are placed in turn on the tree.

We can now produce our ordered list by carrying out a second visit (in-order) search. It's worth noting that the second visit is always made whilst passing *below* the vertex during the navigation of the tree. This makes a quick listing direct from the tree very simple (though, on the other hand, the *formal* definition of 'below' is not simple).

The ordered list is $\langle 5,12,14,15,17,20,22,37 \rangle$

Exercise 7.7.1
Put the list $\langle 96,107,176,43,150,31,66,115 \rangle$ into the form of a binary positional tree, and use the in-order search to produce a sorted list.

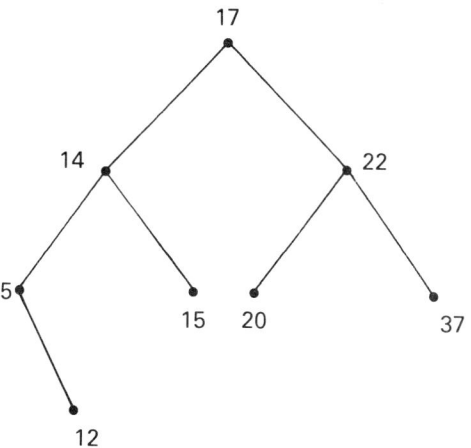

Fig. 7.33.

Exercise 7.7.2
What is the effect of putting an already sorted list into a tree?

Exercise 7.7.3
Suggest a rule for dealing with two elements on the list which have the *same* value (and note that in data processing terms this could correspond, for example, to two different bank accounts which have the *same* account number, which is bad news; or to two cheques paid into the same account, which presumably is good news).

PROBLEMS

1. (a) Show by drawing that the following set of ordered pairs represents a tree,

   $T = \{[a,h],[b,f],[d,j],[h,c],[h,d],[b,g],[a,b],[d,e],[d,i]\}$

   (b) Draw up an action diagram representation.
   (c) Produce a nested bracket version.
2. Look back at the tree representation of the section of the Dewey catalogue in Fig. 7.5. How would you place History of Maths, which has catalogue number 510.9?
3. (a) Draw the tree which is represented in the following nested bracket representation.
   ⟨a⟨b⟨d⟩⟨e⟩⟨f⟩⟩⟨c⟨g⟩⟨h⟨i⟩⟨j⟩⟩⟩⟩
   (b) Show the action diagram representation.
4. (a) Show how the ordered tree shown in Fig. 7.34 can be represented as a binary positional tree.
   (b) Explain the correspondence between the children of a given vertex in the ordered tree, and the representation of the same vertices in the binary positional tree.
   (c) Verify that only one of the three search methods (in-order, pre-oprder or post-order) gives rise to the same visiting sequence in both ordered and binary positional trees.
5. The distances in hundreds of miles between the headquarters of the various outposts of Hannibal's Empire are listed below.

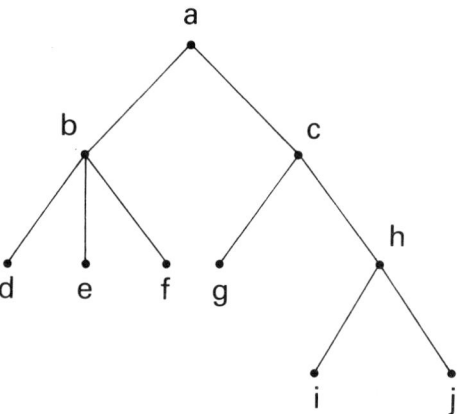

Fig. 7.34.

	Athens	Belgica	Carthage	Dubrovnik	Eastland	Frankfurt
Athens	—	12	9	7	15	10
Belgica	12	—	10	9	3	4
Carthage	9	10	—	11	17	13
Dubrovnik	7	9	11	—	20	14
Eastland	15	3	17	20	—	7
Frankfurt	10	4	13	14	7	—

Use Kruskall's algorithm to minimize the total length of his supply line — that is, a set of routes which connects all six locations. Which two outposts are furthest away from each other along this minimum-length route?

6. (a) A computer network includes six machines, which are connected by a variety of lines (land-line, radio, satellite). The approximate cost of transferring 1 megabyte of data from one machine to another (in either direction) is given in the table below; you may assume that machines can relay data. Use Prim's algorithm to find the minimum-cost network, using vertex a as the initial vertex in the tree.

(b) Assume every machine communicates an equal amount of information to every other machine — relaying if necessary. If this version of the network were adopted as a cost cutting exercise, determine which machine is likely to carry the largest amount of traffic (transmission, reception and relaying).

	A	B	C	D	E	F
A	—	10	5	8	40	12
B	10	—	5	4	—	8
C	5	5	—	6	30	12
D	8	4	6	—	4	20
E	40	—	30	4	—	10
F	12	8	12	20	10	—

(The symbol — means that a given route is not used.)

7. (a) Put the following fully parenthesized expression into an appropriate binary positional tree form.

$$(((a-b)*(c*d))-e)$$

(b) Carry out a pre-order search to produce a prefix expression. Evaluate the prefix expression when $a=3, b=2, c=5, d=10$ and $e=2$. Repeat with a post-order search and evaluate the reverse Polish (postfix) expression.

8. (a) Use the method of Section 7.6 to produce a reverse Polish representation of the expression

$$3+4*(7-5)\uparrow 2$$

and evaluate it.

(b) Draw the tree described in the following doubly linked list.

Address	Record	L. pointer	R. pointer
3	*	7	14
5	/	10	9
7	−	15	17
9	2	0	0
10	12	0	0
12	8	0	0
14	+	12	5
15	5	0	0
17	2	0	0

(c) Assume that the tree represents a fully parenthesized arithmetic expression; show how the *pre-order* search sequence can be used to evaluate the expression.

9. (a) Construct a tree suitable for sorting the given sequence of numbers into an ordered sequence:

21 9 53 5 15 43 62 4 7 12 17 31 49 58 73,

(b) Carry out the appropriate search to obtain the ordered sequence.

(c) Discuss briefly the efficiency of storing records in a tree data structure; why does this particular input give rise to such an efficient tree structure? Devise some input lists which would give rise to less efficient trees.

8
Relations and functions

8.1 WHAT IS A RELATION?

Every graph in Chapters 5 and 6, and every tree in Chapter 7 could be considered as the representation of some relation between objects of entities. Generally a vertex signified an object or entity, and an edge (directed or otherwise) between two vertices indicated some relation between the corresponding entities. The relations were various — b knows a, process p cannot run in parallel with process q, task x must precede task y, and so forth. Now we are going to examine the subject of relations in more detail. We will see that a *function*, as mentioned in the chapter title, is a special type of relation.

We will illustrate the basic ideas about relations by considering two examples, one mathematical, the other more data-relational. We will be invoking some ideas which you first met in Chapter 2, notably about ordered pairs and Cartesian products, so if you are unsure of this material glance back to Chapter 2 before reading any further.

Our first example concerns the relation of a single set H with itself, often called *relation on a set*. Suppose we have a set H where

$$H = \{2,3,4,5,6,7,8\} .$$

We will consider the Cartesian product of H with itself;

$$H \times H = \{[2,2],[2,3], \ldots [8,6],[8,7],[8,8]\} ,$$

which is a set consisting of 49 ordered pairs. A *relation* R, on the set H is simply a *subset* of $H \times H$. (Sometimes R is given the name *binary* relation, which emphasizes the ordered-pair aspect of the description.)

This description becomes more useful when the ordered pairs in the subset obey some kind of 'membership rule'.

What is a relation?

Suppose, for example, that we decide to define a relation D on H by the following membership rule:

$[j,k] \in D$ if k is exactly divisible by j, otherwise $[j,k] \notin D$.

This gives, using the set H defined above:

$D = \{[2,2],[2,4],[2,6],[2,8],[3,3],[3,6],[4,4],[4,8],[5,5],[6,6],[7,7],[8,8]\}$.

There are two conventional ways of notating a binary relation of the form 'j and k are in relation R': the first is $[j,k] \in R$ — 'the ordered pair $[j,k]$ is an element of the relation'; the second is jRk — 'j relates to k'. Similarly we can write the negation 'j and k are not in relation R' as $[j,k] \notin R$, or $j\bar{R}k$, 'j does not relate to k'. So $[2,4] \in D$, whereas $[4,7] \notin D$; and $3D6$, whereas $4\bar{D}6$. You will find that writers differ over which of the two notations to use (some use both interchangeably). We are going to stick to the first representation, which means that the symbol R represents a *set*, used in the other way R represents the verb 'is in relation to'. Although the set version looks a bit unwieldy, it is preferable in many respects to the shorter alternative. You should, however, be able to cope with the other version should you meet it elsewhere.

Example
Classify the following ordered pairs $[x,y]$, either as an element of our relation D, or as an element of $\bar{D}$, or as neither.

$[4,8],[5,8],[1,8]$.

Answer
The ordered pair $[4,8]$ is an element of the relation, so $[4,8] \in D$. The ordered pair $[5,8]$ is an element of $H \times H$, but does not fulfil the rule for membership of D (since 5 into 8 won't go!). So we say that $[5,8] \notin D$.

Our second example of a relation differs from the first in two respects. First, it involves a relation between two distinct sets, whereas the relation D considered above was a relation of the set H with itself. Second, the 'membership rule' for the relation is not defined mathematically, but rather by a practical criterion.

The example concerns a hi-fi dealer who sells four models of CD player and four models of amplifier. The CD players are conveniently named Arc, Bach, Coiner and Disco, so we can represent the set of players by

$C = \{a, b, c, d\}$.

The amplifiers are called Rocker, Stanley, Toner and Universal, giving a set of amplifiers

$A = \{r, s, t, u\}$.

The Cartesian product $C \times A$ will therefore contain 16 ordered pairs, each of the form

[CD player, amplifier].

We now define a relation N as a subset of this Cartesian product containing pairs which satisfy the criterion 'CD player will damage amplifier'. Suppose we ask the dealer to name these pairs for us, and he tells us that the Arc player will damage both the Stanley and Universal amplifiers, while the Disco player will damage the Universal amplifier. We can then write

$$N = \{[a,s], [a,u], [d,u]\} .$$

We can also say that if $[x,y] \in N$ then player x will damage amplifier y.

Notice that in this case the definition of the relation depends on the dealer's judgement — another expert might arrive at a different list of elements of the relation. Notice also that the same information could be conveyed by alternative relations, such as 'will be damaged by' (which is a relation on $A \times C$), or 'is incompatible with' (which is a symmetric relation, and could apply to $A \times C$ or to $C \times A$). This kind of relation may seem more 'vague' than the more mathematical type considered in the first example, but in fact as long as the membership of the relation is unambiguously defined it may be handled in precisely the same way.

We can now extract a general definition of a relation from our two examples. A relation R between sets X and Y is simply a subset of $X \times Y$; if X and Y are identical then R is called a relation on X. In the next section we will investigate ways of representing relations, drawing on ideas about directed graphs and matrices covered earlier in the book.

8.2 DIAGRAMS OF RELATIONS

You may recall from Chapter 2 that a Cartesian product takes its name from Descartes, the mathematician credited amongst other things with the invention of co-ordinate graphs. We saw in Chapter 2 how two related quantities x and y can be represented as a single point $[x,y]$, the x value being measured horizontally from a fixed vertical line, and the y value being measured vertically from a fixed horizontal line. We now make use of this idea to represent a relation, though we do not use continuous x and y *scales*; instead we have certain discrete x and y *values*, which we mark on the axes.

Fig. 8.1 shows how to represent the ordered pairs in our first relation; from the diagram we can see that, for example, $[3,6] \in D$, as there is an entry corresponding to $x = 3$ and $y = 6$. Notice that the absence of an entry indicates that the corresponding ordered pair is *not* in the relation. Notice too that the axes are labelled *only* with values which occur in the ordered pairs of the relation.

Fig. 8.2 makes use of the matrix-like nature of the first diagram to produce a corresponding digraph. Notice that the relation here is *from* elements on the horizontal scale *to* elements on the vertical scale, in the manner of all good Cartesian graphs, though this is the reverse of the convention we used earlier in matrix representations of digraphs.

Fig. 8.3 uses a bipartite graph representation. We tabulate the elements of the set H twice; if we wished we could then make an edge connection from each member of the left-hand list to each member of the right-hand list. This would represent the entire Cartesian product $H \times H$. As we only want to graph the relation D, we join only those pairs which are in D. We will find this form of representation in preferable to the apparently more economical digraph version. We choose to leave out the

Sec. 8.2] Diagrams of relations

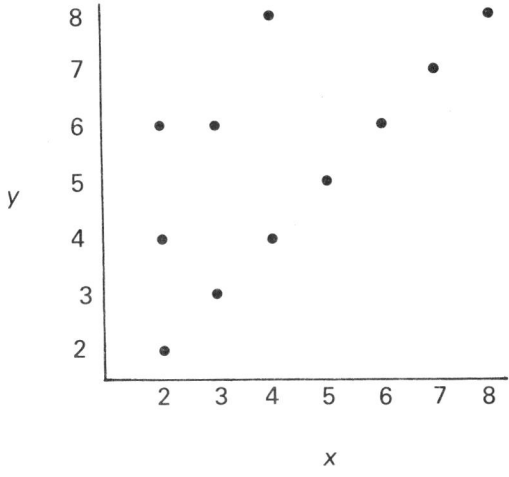

Fig. 8.1.

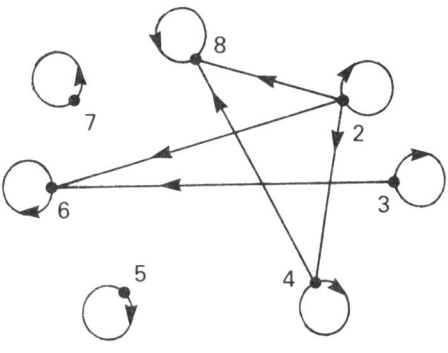

Fig. 8.2.

arrows on the individual edges, instead indicating the left to right convention by means of the single arrow drawn below the graph.

We now move on to relation N, which is shown in two ways (Figs. 8.4 and 8.5). The fact that set A (amplifiers) and set C (CD players) are disjoint, gives rise to Fig. 8.4, which is a Cartesian graph with a distinct set of points on each axis: a, b, c and d on the horizontal axis, $r, s, t,$ and u on the vertical. You should compare the bipartite graph representation which we saw in Fig. 8.3 with the one shown in Fig. 8.5: in the first case, the bipartite graph was on a single set of vertices, which were recorded twice; in this present example, the two sets are disjoint, and the graph is necessarily

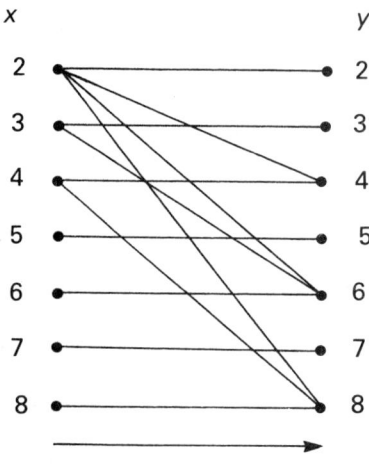

Fig. 8.3.

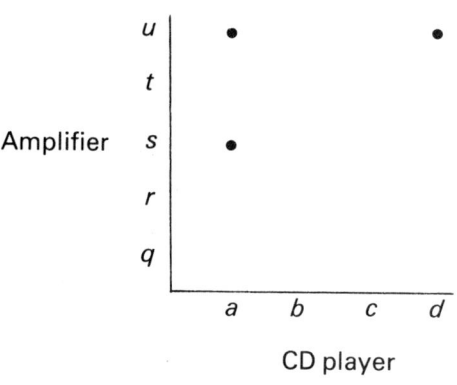

Fig. 8.4.

bipartite. Disjoint sets and their associated bipartite graphs will come in useful later when we apply the ideas of this chapter to relational databases.

8.3 CLASSIFICATION OF RELATIONS ON A SINGLE SET

In this section we will go a little further into the idea of relations on one set. We will begin by examining five properties which such a relation may have, and giving examples of each; we have already come across some of these properties in an informal sort of way. Then we will look at a special type of relation, called an

Sec. 8.3] **Classification of relations on a single set** 199

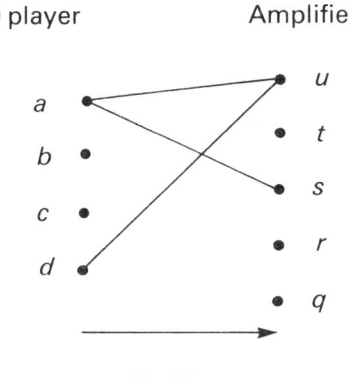

Fig. 8.5.

equivalence relation, which demonstrates several of these properties. Throughout this discussion we will use R to represent a relation on set A, which has members $\{a, b, c, \ldots\}$, and will cite the relations D and N of Section 8.1 as examples.

Property 1
A *reflexive* relation is such that for all $a \in A$, $[a,a] \in R$.

The relation D on set H defined in Section 8.2 was reflexive: $[a,a]$ will belong to D for any $a \in H$ because a always divides itself exactly. Note that this property shows up in Fig. 8.2 as a loop on each vertex, and in Fig. 8.3 as an edge between all pairs of identically labelled vertices.

It is not always easy in the case of a relation defined non-mathematically to decide whether it is reflexive or not; for example, is the relation 'is acquainted with', mentioned in Chapter 5, reflexive?

Property 2
A *symmetric* relation is such that for all $a, b \in A$, $[a,b] \in R \Rightarrow [b,a] \in R$.

The acquaintance relation of Chapter 5 is certainly symmetric: if x is acquainted with y, then presumably y is acquainted with x. However, the relation 'divides exactly' is not symmetric; for example $[2,4] \in D$ but $[4,2] \notin D$.

Property 3
A *transitive* relation is such that for all $a, b, c \in A$, $[a,b] \in R$ and $[b,c] \in R \Rightarrow [a,c] \in R$.

The relation D is transitive, though the fact is perhaps not quite obvious. We need to check that for all cases where $[a,b] \in D$ and $[b,c] \in D$, $[a,c] \in D$ also. D only contains one pair of elements of the form $[a,b]$ and $[b,c]$ — namely, $[2,4]$ and $[4,8]$. So we need to test whether $[2,8]$ is an element of D — and of course it is, because 8 divides exactly by 2. Thus D is a transitive relation.

Notice that it is not easy to establish transitivity by looking at either Fig. 8.1 or

Fig. 8.2. Fig. 8.3, however, gives a clear indication of this property, which could be described thus: a relation is transitive if, in its bipartite graph representation, any pair of vertices connected by a path of length three is also connected by a path of length one.

Property 4
An *asymmetric* relation is such that for all $a, b \in A$, $[a,b] \in R \Rightarrow [b,a] \notin R$.

At first sight it might appear that relation D is asymmetric; certainly if 2 divides 4 then 4 doesn't divide 2. But the definition of asymmetry requires the property to hold for *all* ordered pairs in the relation, whereas in the case of D it will not hold for [2,2], [3,3] and so on. D is therefore *not* asymmetric.

A simple relation which *is* asymmetric is the 'is strictly less than' relation, as you can verify. (Why do we need the word 'strictly' in this statement — in other words, why isn't 'is less than or equal to' an asymmetric relation?)

Property 5
An *antisymmetric* relation is such that for all $a, b \in A$, $[a,b] \in R$ and $[b,a] \in R \Rightarrow a = b$.

This is certainly true for relation D; a divides b exactly *and* b divides a exactly only if a and b are equal. The relation 'is less than or equal to' also has this property, but 'is strictly less than' does not.

The antisymmetric property can be used as a logical device to prove that two things are equal. For example, you can easily show that the relation 'Is a subset of' is antisymmetric, and in Chapter 2 we saw that one way of proving that two sets X and Y are equal was to establish that $X \subseteq Y$ and $Y \subseteq X$.

Example
Suppose the set L consists of a number of straight lines, some of which are parallel. If the line denoted by m is parallel to the line denoted by n, then the relation 'is parallel with' is true and we write $[n,m] \in P$. Use your knowledge of parallelism to check which of the five properties defined above apply to the relation P.

Answer
Reflexive, symmetric, transitive.

Example 2
What about the relation Q 'is perpendicular to'? (Sketching a set of lines some of which are parallel, some perpendicular, may be helpful here.)

Answer
Symmetric.

The relation 'is parallel to' is a particular case of what is called an *equivalence relation*. By definition, an equivalence relation on a set is a relation which is reflexive, symmetric and transitive. In the case of our set of lines, the equivalence relation states that any line is parallel to itself; that if line a is parallel to line b, then line b is parallel to line a; and finally if a is parallel to b and b is parallel to c, then a is parallel to c.

One example of equivalence relations occurs in the context of programming in the following way. Suppose we have a program or process which takes a given string or list as its input, and produces a correctly sorted string as its output. You may have written such programs yourself, using one or more different algorithms. Fig. 8.6

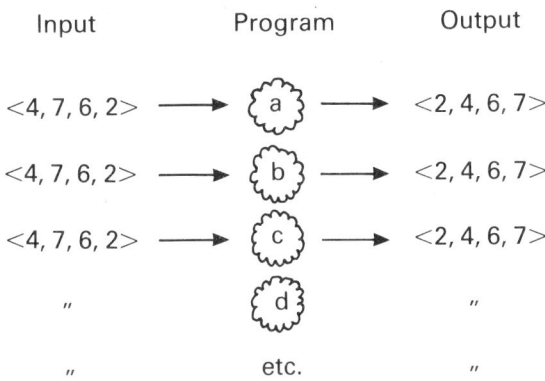

Fig. 8.6.

shows a number of such 'programs'. The cloudiness surrounding the programs a,b,c ... indicates that we aren't actually concerned with how the program works — the machine, the language, the sorting method. It doesn't even matter if one program consists of a reliable clerk with an in-tray and out-tray; as long as an input string $s_1 = \langle 4,7,6,2 \rangle$ is returned as s_0 (the sorted output string $\langle 2,4,6,7 \rangle$) within five minutes, we consider that the program works. At this bargain basement level we say that the programs are *equivalent*.

We can show that this is an equivalence relation on the set P of programs, where $P = \{a,b,c,d,\ldots\}$. The relation, R, can be written as $R = \{[a,b]:[a,b] \in P \times P$ and program a produces the same output as program $b\}$. This means that for the given input string, program a produces the same correctly sorted output string as program b.

The relation is reflexive: $\forall a \in P$. $[a,a] \in R$, since program a always gives the same output as program a.

The relation is symmetric: $\forall a,b \in P$, $[a,b] \in R \Rightarrow [b,a] \in R$ — if a produces the same string as b, then b produces the same string as a.

The relation is transitive: $\forall a,b,c \in P$, $[a,b] \in R \wedge [b,c] \in R \Rightarrow [a,c] \in R$. That is, if a produces the same string as b, and b produces the same string as c, then a produces the same string as c.

To recap — at the risk of seeming to labour the point — all of our programs are *equivalent* with respect to the correctness of processing the string $\langle 4,7,6,2 \rangle$ (though whether they are generally reliable is a different matter). As an exercise, see whether you can apply a similar argument to the problem below, which reverses the roles of program and data.

Example
We have just considered a single input to a number of programs; we can also consider a *number* of inputs to a *single* program. Suppose we consider a set of input strings S; for simplicity we will consider the 24 possible orderings of the input string $\langle 4,7,6,2 \rangle$, that is $S = \{aI, bI, cI, \ldots vI, wI, xI\}$. Note that we have written the elements in list notation. The relation Q on set S is $[a,b] \in Q$, $a,b \in S$. The meaning of the relation Q is that two inputs a and b from set S give the same amount when processed by the program. For instance the input strings $\langle 4,6,2,7 \rangle$ and $\langle 7,2,4,6 \rangle$ should produce the same output $\langle 2,4,6,7 \rangle$. Show that Q is an alternative relation.

Answer
The relation is reflexive, $[a,a] \in Q$: presumably string a will produce the same output as string a (we might consider this as running the data twice through the program). The relation is symmetric, $[a,b] \in Q \Rightarrow [b,a] \in Q$: if string b gives rise to the same output as string a, then string a will give rise to the same output as string b (again we might think of this as running the programs a second time in reverse order). The relation is transitive $[a,b] \in Q \wedge [b,c] \in Q \Rightarrow [a,c] \in Q$.

Again we have an equivalence relation, all inputs are equivalent in terms of the consequent output.

Finally we will conclude our brief examination of equivalence relations with a simple case which will also give you an insight into the apparently commonplace idea of even and odd numbers.

Suppose we define the following relation on a set of integers (or all the integers). *Note*: we are guilty of an inconsistency here! In the context of this chapter, the expression $|x|$ means the modulus of x, or alternatively, the absolute value of x — in other words, the value of x when the sign (+ or −) is ignored. Recall that in set theory it means the cardinality of a set.

$$[x,y] \in R \text{ if, and only if } |x - y| \text{ is even}$$

We can see that

(1) R is reflexive, since $[x,x] \in R$, because $|x - x| = 0$, which is even.
(2) R is symmetrical, since $|x - y|$ and $|y - x|$ are equal, and if $|x - y|$ is even, so is $|y - x|$.

To show that

(3) R is transitive, we note the following: $|x - y|$ is even. So $(x - y)$ is even — if we allow that even numbers can be positive or negative.

We must show that $[x,y] \in R \wedge [y,z] \in R \Rightarrow [x,z] \in R$: to do this we assume that the left-hand side is true, and show that the right-hand side must be true as well.

$(x - y)$ is even, $(y - z)$ is even, so $(x - y) + (y - z)$ is even.
This means that $(x - z)$ is even, which means in turn that $|x - z|$ is even.

This equivalence relation has the effect of *partitioning* any set of integers into two disjoint subsets called *equivalence classes*: one class is the set of odd integers, the other is the set of even integers. But notice that the relation

$$[x,y] \in Q \text{ if and only if } |x - y| \text{ is } odd$$

is *not* and equivalence relation, since it fails the transitivity requirement. (Try $x = 1$, $y = 2$, $z = 3$ to verify this.) One reason is that the odd numbers are odd by virtue of the absence of evenness, whereas all even numbers have a common factor, namely, 2.

Example
What are the equivalence classes for the relation 'is parallel to'?

Example 2
Why is there only one equivalence class for our set of sorting programs?

Answer
Each equivalence class consists of a subset, all of whose elements are parallel lines. Note that an equivalence class may be a subset with a *single* element, as a given line may not have any parallels.

Answer
All the programs produce the same output. Note the following case, however. Suppose the elements of P consist of programs which do *not* all produce the correctly sorted string $\langle 2,4,6,7 \rangle$ — suppose some programs produce, say the reverse string $\langle 7,6,4,2 \rangle$; some produce the original input string; and some produce garbage such as ?*!!!0, which isn't even a string. The relation where $a,b \in P$, and $[a,b] \in R$ when program a produces the *same* output as program b, would give rise to equivalence classes such that all elements of any one equivalence class are programs which produce the same output.

The ideas behind relations have many uses; we will consider two applications in some detail to show how the ideas can be applied to computing. In Section 8.4 we will show a simple example of a relational data base; in Section 8.5 we will consider another special type of relation, called a function.

8.4 COMPOSITION OF RELATIONS

You will recall that a relation may be defined between elements of the same set, or between elements of two distinct sets. We found it convenient to use a bipartite graph representation in both cases; in fact, as we saw, the relation between two disjoint sets gives rise to a graphical representation which is necessarily bipartite. Suppose we have the (abstract) relation R_1, between two distinct sets A and N, where $A = \{a,b,c\}$ $N = \{1,2,3,4\}$ and $R_1 = \{[a,1],[a,2],[b,2],[b,4],[c,3],[c,4]\}$ so that R_1 is a subset of the Cartesian product $A \times N$. We can write this as $R_1 \subset A \times N$. Then consider a second relation R_2, between the set N and another set P where $P = \{w,x,y,z\}$ and $R_2 = \{[1,w],[2,x],[2,y],[4,y],[4,z]\}$; in this case, $R_2 \subset N \times P$ These two relations are shown in bipartite graph form in Fig. 8.7. In Fig. 8.8 we show what is called the

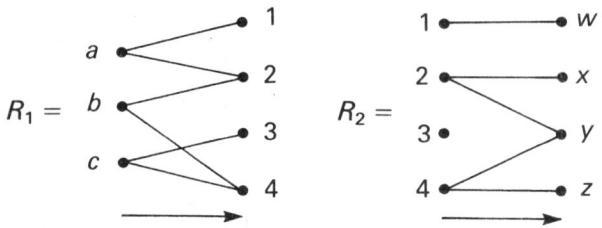

Fig. 8.7.

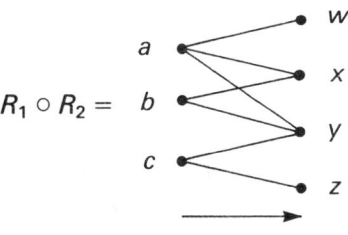

Fig. 8.8.

composition of R_1 with R_2, which is usually written $R_1 \circ R_2$. This composition is a new relation — call it R_c for convenience — between A and P. As with R_1 and R_2, we can show R_c in bipartite graph form, but as you would expect, although the set N contains the links between A and P, it does not feature in the composition.

Example
(a) Write out the set of ordered pairs of the composite relation R_c where $R_c = R_1 \circ R_2$, and $R_1 \subset A \times B$ and $R_2 \subset B \times C$ are defined as follows:

$A = \{a,b,c\}$ $B = \{5,6,7,8\}$ $C = \{w,x\}$
$R_1 = \{[a,5],[a,7],[b,7],[c,8]\}$
$R_2 = \{[5,x],[6,w],[7,x],[8,w]\}$

(b) Draw the bipartite graph representation of R_1, R_2 and R_c.

Answer
(a) $R_c = \{[a,x],[b,x],[c,w]\}$
(b) See Fig. 8.9.

Now although it is often good practice to simplify information — and the bipartite graph of the composition R_c looks reasonably simple — we have lost some of the original information which R_1 and R_2 contained, and some of that information

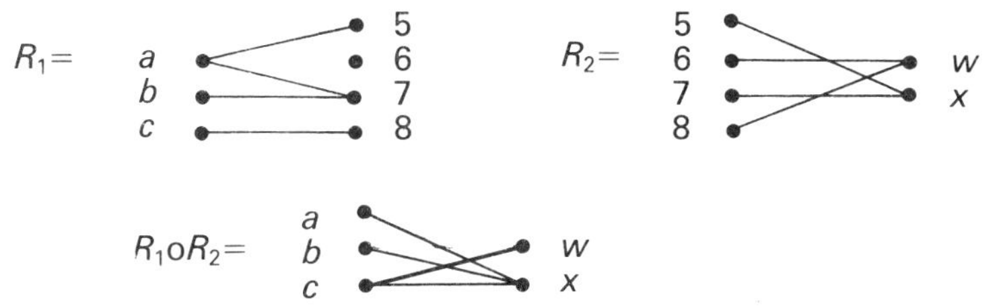

Sec. 8.4]   **Composition of relations**   205

$R_1 =$ a, b, c → 5, 6, 7, 8 $R_2 =$ 5, 6, 7, 8 → w, x

$R_1 \circ R_2 =$ a, b, c → w, x

Fig. 8.9.

may be important. Certainly, for information processing purposes, we ought to retain the link, (or *key*) which connected corresponding elements of A and P in the first place. Notice that an element in the first set of a composite relation may relate to an element in the third set through more than one key, as happened in the exercise above; for example, the ordered pair [a,x] occurs in the composition because of the existence of the ordered pairs [a,5] and [5,x]: but it occurs also because of the ordered pairs [a,7] and [7, x]. (You should be able to find a similar duplication in the example illustrated by Figs. 8.7 and 8.8, earlier in the section.) For this reason a diagram may be useful for some purposes, but we won't rely on it. It's obviously possible to avoid this problem by representing the composition of two relations as a set of ordered *triples*, but in fact this representation is more conveniently shown in the form of a table.

Example
(a) Write out the ordered triples of the composition shown in Figs 8.7 and 8.8.
(b) Draw up a table to represent the composition.

Answer
(a) The ordered triples are $R_c =$ {[a,1,w], [a,2,x], [a,2,y], [b,2,x], [b,2,y], [b,4,y], [b,4,z], [c,4,y], [c,4,z]}
(b) See Fig. 8.10.

Now we will revisit the hi-fi problem introduced in Section 8.1, and show how a composition might be interpreted in practical terms. Then we will be in a position to take a brief look at the idea of a relational database. On the way we will be revising tuples, remembering logical connectives, and extending our ideas about matrix representation. However, all the basic ideas are simple ones!

Suppose that our hi-fi dealer of Section 8.1 takes a positive approach to the CD players and amplifiers. Instead of applying the relation N — where [a,s]∈N meant that the Arc CD player was incompatible with the Stanley amplifier — he decides to apply the relation M, meaning 'matches' or 'is compatible with' — for example, [a,r]∈M indicates that the Arc CD player is compatible with the Rocker amplifier. We can also show the set of compatible relations in bipartite graph form, as in Fig. 8.11, though as we suggested earlier the tabular form shown in Fig. 8.12 will turn

Set A	Set N	Set P
a	1	w
a	2	x
a	2	y
b	2	x
b	2	y
b	4	y
b	4	z
c	4	y
c	4	z

Fig. 8.10.

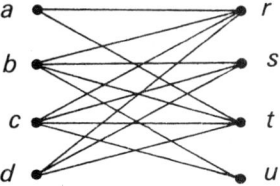

Fig. 8.11.

out to be more seviceable. The ordered-pair form, as shown below is another possibility.

$$M = \{[a,r],[a,t],[b,r],[b,s],[b,t],[b,u],[c,r],[c,s],[c,t],[c,u],[d,r],[d,s],[d,t]\}$$

Whichever form of representation we use, we can describe the information in database terminology as well. In database terms, we can consider M as a *file* which contains nine *records*. Each record consists of just two *fields*. Each field has a *name*: the name of the first field is 'CD player', and the name of the second field is 'amplifier'. For a given record to represent anything, the fields in the record must have specific *values*.

Notice that in this example both the fields of M are alphabetic and contain alphabetic values; in other cases we will find fields specified as True/False, numeric and so on. The corresponding values will be alphabetic, True/False, or numerical also. Don't forget that a name is as much a variable as is a number.

Our ordered pair/table form may seem to be an unnecessarily repetitive way of

CD player	Amplifier
a	r
a	t
b	r
b	s
b	t
b	u
c	r
c	s
c	t
c	u
d	r
d	s
d	t

Fig. 8.12.

recording the information: why not write out each CD name and attach the names of all compatible amplifiers (though even then there would still be a certain amount of repetition)? We won't try to answer this question, but we must realize that any precise information will take up a certain amount of space. More importantly, our chosen form means that all records have the same structure and size, and this form is well suited to formal mathematical treatment — and by implication, to computer implementation.

When we have a clearly specified file like this, we can consider it as a very simple database, and we can imagine it as being stored in some accessible computerized form. One way to address enquiries to and obtain information from the database is by means of what is called a *query language*: although there are other ways of interrogating a data base, and although query languages differ between different systems, we can see the principles from the following example. Suppose the dealer wants to give a list of amplifiers which are compatible with both the Arc and the Bach. He could then use the following sequence of commands (the underlined words are words of the query language):

<u>FROM</u>   M
<u>SELECT</u>   amplifier
<u>WHERE</u>   CD = Arc <u>OR</u> CD = Bach

The first line tells the machine which file (or files) are to be interrogated. The second line indicates the name of the field (or fields) which are to be output — in this case just the amplifier name. The third line supplies the selection criteria, using the logical connective OR.

The output will be 'Rocker, Toner, Rocker, Stanley, Toner, Universal', corresponding to the six ordered pairs which have either a or b as the first element. Most systems would give the output in table form with the heading 'amplifier'. If you have any doubts about the reason for the repeated names, refer back to Fig. 8.12.

Now we will show how a data base can be made up of several related files. We will do this by expanding the hi-fi business into the selling of loudspeakers; suppose there are two types, the Jazz and the King, whose names are given by initial in the set L, so that $L = \{j,k\}$. Further, suppose the dealer considers that only certain amplifier/loudspeaker combinations are satisfactory, irrespective of the CD player to be used. His set S, of compatible pairs looks like this (the tabular form is shown in Fig. 8.13):

$$S = \{[r,j],[r,k],[s,k],[t,k],[u,k]\}$$

Amplifier	Speaker
r	j
r	k
s	k
t	k
u	k

Fig. 8.13.

We can apply the idea of a composition to the relations S and M and get the relation in ordered pair form

$$M \circ S = \{[a,j],[a,k],[b,j],[b,k],[d,j],[d,k]\} \ .$$

In bipartite graph form, this turns out to be a *complete* bipartite graph, with the simple interpretation that each CD player is compatible with every speaker.

As Fig. 8.14 shows, the tabular form corresponding to ordered *triples* is much more informative, and indicates all compatibilities. (The unwieldy 'tripartite' graph of the composition is shown in Fig. 8.15, but is not recommended as a representation). Note that even in a small case like this, these are 32 ($= 4 \times 4 \times 2$) ordered triples in the extended Cartesian product $C \times A \times L$ and there are 17 ordered triples in the composition relation. This is a much higher 'hit rate' than we would be likely to meet in practice: in life-sized examples the Cartesian product might consist of many millions of ordered tuples, with only a small fraction of such tuples belonging to a given relation.

Our query language can be extended to make enquiries about the joint file M

Sec. 8.4]   **Composition of relations**   209

CD player	Amplifier	Speaker
a	r	j
a	r	k
a	t	k
b	r	j
b	r	k
b	s	k
b	t	k
b	u	k
c	r	j
c	r	k
c	s	k
c	t	k
c	u	k
d	r	j
d	r	k
d	s	k
d	t	k

Fig. 8.14.

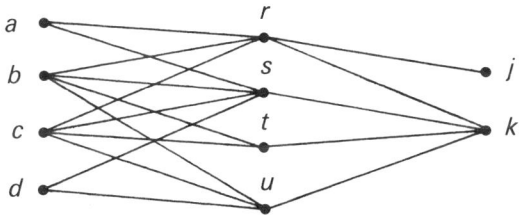

Fig. 8.15.

combined with *S*. Suppose the dealer wants a list of all compatible CD player/amplifier/speaker combinations. He can make the query as follows

  <u>FROM</u> *M*, *S*
  <u>SELECT</u> CD, amplifier, speaker
  <u>WHERE</u> amplifier <u>IN</u> *M* = amplifier *IN S*

The two files have a common key, which is 'amplifier', and it is this *relation* between the files which enables us to select appropriately. If the value of the amplifier is the same for two records — one in file M and the other in file S — then we can *join* the two records and select values from any of the three fields in the joint record (in this example, we selected all three fields). If we had selected just CD and speaker, there would have been repeated entries, such as $[a,k]$, which would have been selected twice — once as a member of the triple $[a,r,k]$, and once as a member of $[a,t,k]$.

Exercise 8.4.1
The relation between four named students and their first-level subjects is given in the table below: an entry of 1 (TRUE) corresponds to the fact that the ordered pair (student, first-level subject) is an element of the relation R_1; in set notation we could write

$$R_1 = \{[x,y]: \text{student } x \text{ is studying first-level subject } y\}$$

Student	First-level subject			
	Bio.	Chem.	Maths	Physics
Smith	0	1	0	1
Jones	1	0	1	0
Dylan	1	1	0	0
Marx	0	0	0	1

Also in set notation
$R_2 = \{[a,b]$: first-level subject a is a pre-requisite for second-level subject b)
and this information is shown in the table below.

Second level	First-level subject			
	Bio.	Chem.	Maths	Physics
Ecology	1	0	0	1
Modelling	0	0	1	0
Natural sc.	0	1	0	1

(a) Represent these two relations in bipartite graph form.
(b) Find the composition $R_1 \circ R_2$ and represent it in bipartite graph form; in matrix form; as a set of ordered pairs; and finally as a set of ordered *triples*.

(c) What does the composition represent in terms of information?

8.5 THE IDEA OF A FUNCTION

We are now going to consider three examples of relations which turn out to have a rather important special feature in common — so important that relations having this feature are distinguished by the name of *functions*.

Example 1 we have already met in Section 8.3, where we imagined a program which could sort a four-digit string into an ordered string. There are actually 24 different orderings of four digits (we will see why in Chapter 9), but a valid sorting program will always produce the *same* output string from any input string containing the same four digits.

Example 2 concerns a skilled arithmetician, who is given a calendar date such as 2/9/1988, and can then calculate the day of the week corresponding to that date.

Example 3 involves no arithmetic: if a student at a college is named, we can then determine which year of his or her course the student is on.

Each of these three examples is clearly a relation. But they have a more specific feature in common, which distinguishes them as *functions*. In Example 1, all 24 possible input strings consisting of the same four digits are in relation to just one output string. In Example 2 any given calendar date is in relation to just one of the days of the week; and in Example 3 each student is in relation to just one of the four course years 1, 2, 3 or 4.

We can get more insight into the special nature of these relations by considering some ordered pairs and asking whether they could, or could not, belong to the corresponding relation.

For Example 1, a typical ordered pair might be $[\langle 2,7,6,4 \rangle, \langle 2,4,6,7 \rangle]$ — that is, the string $\langle 2,4,6,7 \rangle$ is the sorted version of $\langle 2,7,6,4 \rangle$. $[\langle 7,4,6,2 \rangle, \langle 2,4,6,7 \rangle]$ is also a valid possibility. The fact that the right-hand element of both ordered pairs is the same does not matter — it simply reflects the fact that many different strings produce the same sorted string. On the other hand, $[\langle 2,7,6,4 \rangle, \langle 2,4,7,6 \rangle]$ *cannot* be a valid member of the relation; it it were, there would be two different sorted orders for the same string, which is clearly not possible. So two ordered pairs with the same left-hand element indicate that something has gone wrong.

In the same way, [2/9/1988, Friday] and [20/10/1988, Friday] are both valid members of the relation in Example 2 — both have the same right-hand element because there have been many Fridays in history. But [2/9/1988, Thursday] could *not* be valid, since it would suggest that the same day could simultaneously be a Thursday and a Friday. Again, the individual left-hand elements indicate the problem.

Finally, [Horace Morris, 3] and [Nathan Daboll, 3] are acceptable as members of the Example 3 relation — two students can be in the same year — but [Horace Morris, 2] would be inconsistent with [Horace Morris, 3] — the same student cannot be in two different years.

To generalize, what we have observed in all three relations is that, given the first element of an ordered pair which belongs to the relation, the second element is then *uniquely* defined. This is one of the characteristics which makes these three relations into *functions*. A function may be 'many-to-one', as with the examples here, where many dates relate to the same day of the week, many students to the same course

212   Relations and functions   [Ch. 8

year. It could be 'one-to-one', as with the mathematically defined relation 'is the square of' (try this with some numbers if you cannot see it directly). What it *cannot* be is 'one-to-many' — the invalid cases cited above illustrate that.

The many-to-one feature of our three examples, which we will now call respectively F_1, F_2 and F_3, stands out clearly in their bipartite graph representations. Partial bipartite graphs for E_1, F_2 and F_3 are shown in Fig. 8.16; we can't show that the full

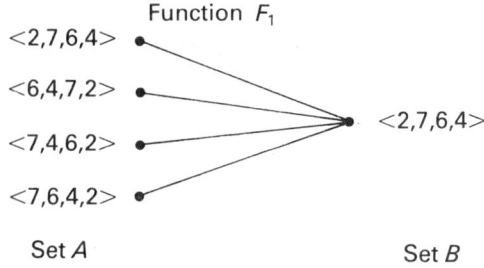

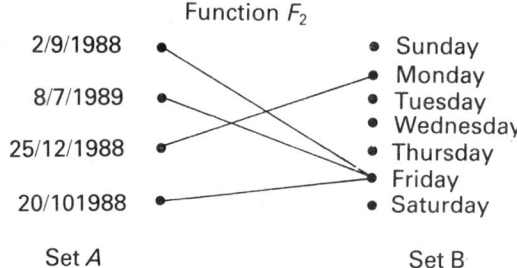

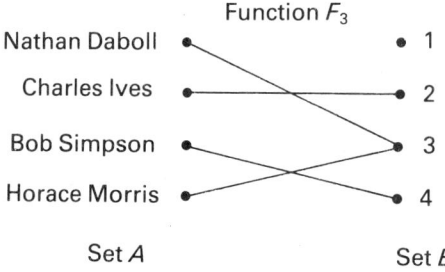

Fig. 8.16.

graphs for these functions because we have not specified all the ordered pairs involved. Nevertheless the fact that elements of the first set correspond to a unique element of the second set should be clear.

In order to write down a more general definition of a function, we need to introduce some terminology. We refer to the elements of the left-hand sets in Fig. 8.16 as *objects*, and to those of the right-hand set as *images*. We will also refer to the set of all objects for a particular function as its *source set*, and to the set of possible images as the *target set*. It is convenient to refer to these two sets as S and T respectively, and to their elements as s and t. The second defining property of a function can then be summarized by saying that the function operates on *all* elements of the source set — there is no object which does not have an image (though, as we will see, there may be members of the target set which are not images of any s in the source set).

We can now formally define a function F; our definition will have two parts:

(1) For all $s \in S \, \exists t \in T$ such that $[s,t] \in F$.
   In words, this expresses the fact that for *every* object s there exists an image t — the function operates on the whole of the source set.
(2) Any given s has a *unique* image t. To express this formally, we need to rephrase it slightly: if a given s apparently has two images t_1 and t_2, then $t_1 = t_2$.
   So for all $s \in S$, $t_1, t_2 \in T$, $[s,t_1] \in F$ and $[s,t_2] \in F \Rightarrow t_1 = t_2$.

For a relation to be a function, both parts of this definition must be shown to be true. Notice that neither part of the definition precludes the possibility that many objects map to the same image.

Example
Fig. 8.17 shows four different relations between sets $S = \{a,b,c,d\}$ and $T = \{w,x,y,z\}$. Identify the relations which are also functions, and in the case of those which are not functions, state whether parts 1 or 2 of the definition (or both) are not satisfied.

Answer
Fig. 8.17(a) and (d) are functions.
   Fig. 8.17(b) fails in part 1 of the definition, since there is no image corresponding to d in the source set.
   Fig. 8.17(c) passes on part 1, but fails on part 2 of the definition — the element d in the source set *maps* (as we say) to both y and z in the target set, so we do not have a unique image for every object.
   It is worth noting that the function in Figs. 8.17(d) has a further special property: it is *one-to-one*, so that every image arises from a unique object. We can therefore say that the reverse relation, which takes us back from T to S, is also a function. This would not be the case with Fig. 8.17(a). We call such a function *invertible*, and refer to the function from T to S as the *inverse function*.

8.6 FUNCTIONS AND PROGRAMS

We have seen that certain relations can be classed as functions; now we will go one short step further and consider briefly how some functions are closely connected to *programs*, or procedures. Consider the three examples of the last section in turn: the first was an example of a string input followed by a string output; the second was a

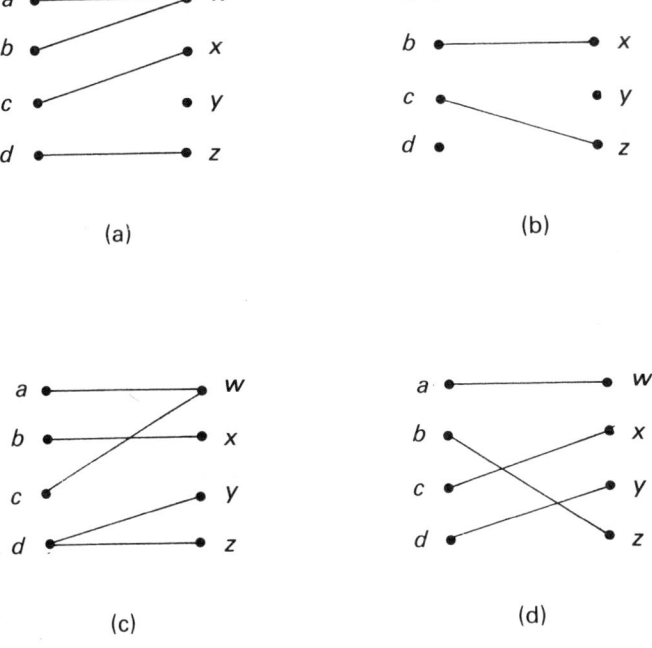

Fig. 8.17.

three number input followed by a single word output; and the third was a name input followed by a single digit output.

In each case we have some kind of *process*; algorithmic in the case of the sorter; some form of 'look-up table' in the student and year case; and, as it turns out, either a large calendar (as a look-up table) or an algorithm in the 'day of the date' case. We'll take this last-mentioned case, and use a rule given by the nineteenth century mathematician Zeller, to illustrate the functional nature of a program.

Zeller's rule
The rule may be used to find the day of the week of any calendar date from 1 A.D. onwards. In English-speaking countries two calendars have been used: the Julian calendar, from the time of Julius Caesar to 2 Sept. 1752; and the Gregorian calendar, which has been in use since 14 Sept. 1752. The steps are follows:

(1) Calculate the 'golden number' G for the given year y, thus:

for the Julian calendar $G = 0$;
for the Gregorian calendar $G = \text{int}(y/100) - \text{int}(y/400) - 2$.
(Note that int(x) returns the integer part of x; in other words, it is itself a function.)

(2) Suppose the given calendar date is day 'd', month 'm' and year 'y', and that we

wish to find the day of the week 'r', defined so that if $r = 0$ the day is Saturday, if $r = 1$ the day is Sunday and so on through the week to Friday, for which $r = 6$.

To calculate r we use the formula

$$r = (d + 2m + \text{int}(3(m + 1)/5) + y + \text{int}(y/4) - G) \text{ (modulo 7)}.$$

(Note that January and February must be treated as the 13th and 14th months of the previous year.)

This may not look like a program, but the action diagram below is a more formalized version, which could easily be turned into a program.

Zeller's rule (with the year 1752 avoided!)

```
⎡ENTRY
│   y (year)
│   m (month)
│   d (day)
│   ⎡IF y⟨⟩1752
│   │       ⎡IF y⟨1752
│   │       │   G = 0
│   │       ⊢ELSE
│   │       │   G = int(y/100) − int(y/400)
│   │       ⎣ENDIF
│   │       ⎡IF m⟨3
│   │       │   m = m + 12
│   │       │   y = y − 1
│   │       ⎣ENDIF
│   │       r = (d + 2∗m + int(3∗(m + 1)/5) + y + int(y/4) − G) (modulo 7)
│   │       ⎡IF r = 0, 1, 2, 3, 4, 5, 6
│   │       │   print 'Saturday', 'Sunday', 'Monday', 'Tuesday', 'Wednesday',
│   │       │   'Thursday', 'Friday'
│   │       ⎣ENDIF
│   ⊢ELSE
│   │   print 'No answer'
│   ⎣ENDIF
⎣EXIT
```

As far as the rather sparse bipartite graph of this function is concerned (Fig. 8.16), we can append as many left-hand vertices (dates in history, that is) to the graph as we like, and by means of the rule-/action diagram/program we can find the appropriate right-hand vertex (the day of the week). Notice that the input is in fact an ordered triple of the form (day, month, year) and that the output is an alphabetical variable with a value corresponding to one of the days of the week.

As it stands, the action diagram represents a far from perfect program. In particular it will accept dates which don't correspond to valid calendar dates. As an example, you could input a date [31,2,1989] and you would get the output Tuesday: so we see that what are called the *pre-conditions* for the program are not completely defined. The program may give an apparently valid answer to an invalid question. A more sophisticated version of the program would contain checks on the range and validity of an input: for example range checks for day (between 1 and 31), for month (between 1 and 12), and for year (between 0 and 2000) would rule out some invalid inputs — though [31,2,1989] would still be Tuesday! Another pre-condition is that all the elements of the input array should be integer — though we could argue that the pre-condition of a valid date covers this requirement, since all valid dates consist of integers.

If we examine what are called the *post-conditions*, that is, the set of valid *outputs*, things look a bit brighter. We see that *if* the program gives an output, it will be a valid output — that is, the name of one of the days of the week. So we can say that the *post-conditions* of the program are satisfied; and while the result may be for a non-existent date, or Zeller's rule may contain an error, giving a wrong answer, the answer as such will still be consistent with the post-conditions.

This very brief look at the requirements for a program to be valid (or reliable or safe!) is just a starting point for the serious business of analysis and validation of computer programs by means of mathematics. As a further step we would need to verify that the value of r returned by the program is bound to be an integer: then we need to show that its value must be in the range 0 to 7 — so there are many other conditions which must be met. In the long term — and this is by no means the case at the time of writing — the mathematics of this type of *software engineering* will be as specific and formal as the mathematics of mechanical and electrical engineering are today.

Example
Describe pre- and post-conditions for the program relating students to course year.

Answer
In this case the pre-conditions are that the input should consist of the first name of student in the college, followed by the second name of the same student. The post-condition is that the output should be an integer number between 1 and 4.

PROBLEMS
1(a) Consider the relation R defined as follows:

Let $A = \{3,5,7,10\}$, and $B = \{4,6,7\}$.
$[a,b] \in R$ if and only if $a \in A$, $b \in B$, and $|a - b|$ is even.

Represent this relation as a subset of the Cartesian product $A \times B$.

(b) Draw a bipartite graph representation of relation R.

(c) Now consider the more general relation S_n defined on the set I of all integers thus:

$[a,b] \in S_n$ if and only if $a,b \in I$ and $|a - b|$ divides exactly by the integer n.

Either show that S is an equivalence relation, *or* determine on which of the equivalence relation criteria it fails.

2. Define sets $C = \{a,b,c,d\}$, $D = \{w,x,y,z\}$ and $E = \{p,q,r\}$.
Four relations are defined as follows:

$R = \{[a,w],[b,w],[c,y],[d,z]\}$, so $R \subset C \times D$.
$S = \{[a,w],[b,y],[c,z],[d,x]\}$, so $S \subset C \times D$.
$T = \{[a,p],[b,q],[c,q],[d,q]\}$, so $T \subset C \times E$.

$U = \{[w,p],[x,q],[x,r],[y,q]\}$, so $U \subset D \times E$.

(a) Draw the bipartite graph representation of these relations.
(b) Which of these relations are functions?
(c) In which cases is the inverse relation also a function?

3. Sets I, P and Q are fields in a simple database;

$I = \{\text{cup, saucer, spoon}\}$, representing the very limited range of items sold by a shop.
$P = \{10, 15, 12, 8, 17\}$, representing a set of unit prices in pence.
$Q = \{x : x \text{ is an integer less than } 100\}$, representing a set of possible purchase quantities.

The relation $R_1 \subset P \times I$ may be considered as a current price file:
$R_1 = \{[15, \text{cup}], [12, \text{saucer}], [8, \text{spoon}]\}$.

The relation $R_2 \subset I \times Q$ represents a particular customer's order:
$R_2 = \{[\text{cup}, 6], [\text{saucer}, 4], [\text{spoon}, 3]\}$.

(a) Draw up bipartite graph representations of the two relations, and of the composition $R_1 \circ R_2$.
(b) Compare the ordered triples of the composition with the bipartite graph representation.
(c) If both R_1 and R_2 are considered as computer files, write our a set of query language instructions which produce a set of data equivalent to the composition $R_1 \circ R_2$.

4. A restaurant with a small menu serves fish, chips and peas. A customer's order may consist of nothing or of any combination of fish, chips or peas — so a customer's order is a subset of the set A, where $A = \{f,c,p\}$. There is a fixed service charge of £1 to discourage people from ordering nothing: fish costs £2; chips £0.50 and peas £0.50.

(a) Write out the power set of A; call this set M.
(b) Write out the elements of the set B, which is a set of prices, defined by $B = \{x : y \in N \text{ and } y < 9 \text{ and } x = 0.5y\}$
(c) Three students each order a meal: Gerry — who is a vegetarian — orders peas; Horace orders the lot; and Ian orders just chips. For convenience consider the students' names to be elements of the set S, where $S = \{g,h,i\}$. The relation R_1 can then be described as follows.

$R_1 = \{[x,y] : [x,y] \in S \times M \text{ and } x \text{ orders meal } y\}$

Draw a bipartite graph representation of this relation.
(d) Describe (in similar terms to the relation R_1) the relation R_2, which maps meals (that is, elements of M) onto prices; and show it as a bipartite graph.
(e) Tabulate the composition $R_1 \circ R_2$ in three columns.
(f) Which, if any of R_1, R_2 and the composition $R_1 \circ R_2$ are also functions?

5. The following cities are accessible from the named motorways: Bath M4; Birmingham M5, M6; Winchester M3; Bristol M4, M5; Canterbury M2; Cardiff M4; Exeter M5; Leeds M1; London M1, M2, M3, M4; Manchester M6; Sheffield M1. Suppose that set C consists of the names of the cities and that the set M consists of the names of the motorways as given above. The relation A can be defined as follows.

$A = \{[x,y] : [x,y] \in M \times C \text{ and motorway } x \text{ gives access to city } y\}$

(a) How many elements are there in the Cartesian product $M \times C$?
How many elements are there in A?
(b) Tabulate the ordered pairs which are elements of the set D, defined thus:

$D \subset A$ and $[x,y] \in D$ if $[x,y] \in A$ and $x = $ 'M4' or $y = $ 'London'.

(c) The cities of Bath, Winchester, Canterbury and London are classified as historic cities: the rest are classified as not-historic. Suppose we have the Boolean set $H = \{T,F\}$, where a value T corresponds to historic, and a value F corresponds to not-historic. Draw a bipartite graph representation of the relation B, where

$B = \{[y,z]:[y,z] \in C \times H$ and city y has historic classification $z\}$

(d) How many elements does the Cartesian product $C \times H$ have?
(e) How many elements does the composition $A \circ B$ have?
(f) Suppose $A \circ B = K$: write out the set of ordered triples V, defined below:

$V = \{[x,y,z]: [x,y,z] \in K$ and $x = $ 'M4' and $z = T\}$

(g) Write out a set definition in terms of K, similar to the one in (f), of the names of all historic cities.

6. (a) Suppose that R is the set of current registration numbers of private motor vehicles in the UK, O is the set of owners of private motor vehicles in the UK, and M is the set of names of makes of private motor vehicles in the UK. Which of the following relations are (likely to be) functions?

$R_1 = \{[x,y]: [x,y] \in O \times R$ and the person named x owns the vehicle with registration $y\}$

$R_2 = \{[x,y]: [x,y] \in R \times O$ and the vehicle with registration x is owned by the person named $y\}$

$R_3 = \{[x,y]: [x,y] \in R \times M$ and the vehicle with registration x is of make $y\}$

(b) What does the composition $R_1 = \{[x,y]: [x,y] \in R_1 \circ R_3$ represent?

7. Describe some of the preconditions for the input to the following programs or calculations. What are plausible post-conditions?

(a) To calculate the mileage of a given trip from the start and finish mileometer readings (mileometers have five digits on their displays).
(b) To find the square root of an integer number.
(c) Given three straight lines of lengths a, b, and c respectively; to calculate the angles of the triangle formed by the three complete lines.

9
Counting and probability

9.1 NEW WAYS OF COUNTING

We tend to think of the ability to count as something pretty basic, on a level with being able to read; saying to someone, 'Can't you count?' carries the implication, 'Well, you must be fairly dim, then'. However, if we think in a little more detail about what counting actually means, especially in the light of some of the counting problems we have already met earlier in the book, we find it is not quite such a simple process as we might imagine.

Back in Chapter 2, we saw how the child's method of counting on its fingers really involves putting the things being counted in a one-to-one correspondence with a set, the number of elements in which is known — namely, the fingers. Once we are confident enough in the number system, we do away with the intermediate set of fingers, and can count larger sets of items, by putting them directly into one-to-one correspondence with the known set of natural numbers.

Later still, we may feel that in certain situations we can dispense with some of the intervening numbers and count 'ten, twenty, thirty, . . .' or 'One hundred, two hundred, . . .'. Sometimes, in fact, this is unavoidable — if we want to count items which are packaged in boxes of 100, we may not be willing to empty each box, so we simply count the boxes and multiply by 100 to obtain the number of items — trusting that there are none missing! But however large the 'counting units' may be, we are still in some way scanning the entire set of things to be counted.

This kind of direct counting is called enumeration, and it works quite well in many cases — indeed, often it is the only method at our disposal. If you are taking an inventory in a warehouse, for example, it may well be necessary physically to count all the items of each type on the shelves.

Unfortunately many counting problems are extremely cumbersome if approached in this enumerative fashion. In Chapter 2 we needed a way of counting the subsets of a set with k elements; in Chapter 5 we wanted to know how many edges a complete graph with k vertices would have ... no doubt you could add further examples to this list. In all these cases, a direct enumeration of the possibilities is time-consuming. In the case of the subsets and the complete graph, we would need first to carry out an enumeration for several different values of k, then attempt to generalize the result, and finally prove our conjectured result true, probably by an induction proof (see Chapter 3). This is the sort of complication which makes one feel that 'there must be a quicker way'!

The feeling that a quicker way must exist, which we might call a sort of constructive laziness, has led to many important developments in mathematics — first-rate mathematicians are never the people who are happy to wade through a mass of calculations, but always the ones who want to find how to reach the same conclusion in two lines. In the case of counting, the urge to circumvent enumeration has led to the development of a whole area of mathematics known as combinatorics, and some elementary results from that area will occupy us for the first half of this chapter.

Before we start looking at the ideas formally, you may like to set yourself thinking in the right way by considering the following problem. You have decided to buy a personal computer, and have narrowed down the choice to just three possible models. You also need a monitor, and there are only two suitable models on the market. Finally, you need to link the system to any of four models of printer. How many possible configurations are there for the system? (Think back to the hi-fi dealer's problem in Chapter 8 for a way of approaching this question.)

9.2 'MENU-TYPE' PROBLEMS — THE MULTIPLICATION PRINCIPLE

The answer to the question posed above should be 24. Of course, it is possible to enumerate all 24 possibilities here without too much trouble: if the computers are denoted by A, B, C, the monitors by 1, 2, and the printers by W, X, Y, Z, then we could list the various configurations thus:

$$A1W, A1X, A1Y, A1Z, A2W, \ldots, C2Z.$$

(Notice that if you are resorting to enumeration, it's important to be systematic about it, otherwise cases can easily be missed — or included more than once.)

However, we hope that you did not use direct enumeration to reach the answer. There's an opportunity here to use some of the ideas of Chapter 7, and show the different configurations by means of a tree structure, as seen in Fig. 9.1. Alternatively, we could show the situation by an extension of the bipartite graph idea, as in Fig. 9.2. Then the required number of configurations is given by the number of paths from the root of the tree to a terminal vertex, or the number of routes leading from a member of the set of computers, via a member of the set of monitors, to a member of the set of printers.

This makes it clear that the possible configurations are actually ordered triples of the form (computer, monitor, printer). The total set of possibilities is therefore the Cartesian product

$$\{\text{all computers}\} \times \{\text{all monitors}\} \times \{\text{all printers}\}.$$

Sec. 9.2]   'Menu-type' problems — the multiplication principle   221

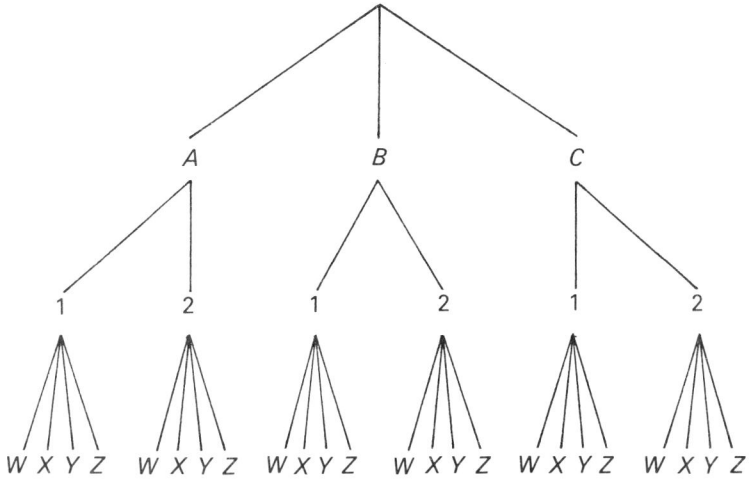

Fig. 9.1.

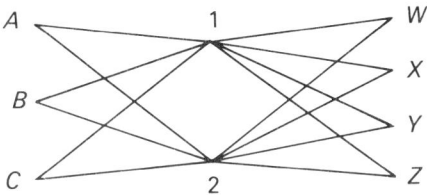

Fig. 9.2.

It is probably easiest to spot the 'shortcut' method by looking at Fig. 9.1. Each of the three initial branches, representing the three computers, splits into two at the second stage, showing the two types of monitor. So at this stage we have $3 \times 2 = 6$ possible combinations, any of which could be followed by any of the four printers. Thus the total number of possibilities is $3 \times 2 \times 4 = 24$.

What we are saying here is that if we have a sequence in which the first item can be selected in n_1 ways, the second in n_2 ways, and so on, then the total number of possibilities is $n_1 \times n_2 \times \ldots$. We have called the kind of problem where this multiplication principle applies a 'menu-type' problem, because this is exactly what happens when you select a meal from a menu giving a choice of, say, starters, main courses and sweets. The total number of possible different meals you could order (regarding 'a meal' as the totality of the three courses) would then be: number of starters × number of main courses × number of sweets.

Another way of remembering the circumstances under which the multiplication principle applies is to think of it as the 'and' rule. It is valid in situations where we are choosing one out of n_1 items *and* one out of n_2 items and so on.

Here are a couple more examples to show how the principle works.

Example 1
Generalizing what was said above, we note that the Cartesian product of sets $A_1, A_2, A_3, \ldots$, written as $A_1 \times A_2 \times A_3 \times \ldots \times A_n$, is defined as the set of all n-tuples $[x_1, x_2, x_3, \ldots, x_n]$ such that $x_1 \in A_1, x_2 \in A_2, \ldots$. Because there are $|A_k|$ elements in A_k where $|A_k|$ denotes the number of elements in set A_k, there are $|A_k|$ ways of choosing x_k. So the multiplication principle tells us that

$$|A_1 \times A_2 \times \ldots \times A_n| = |A_1| \times |A_2| \times \ldots \times |A_n|$$

since we are taking an element from A_1 *and* one from A_2 and so on.

Example 2
A student can choose from six essay topics for a particular assignment; if she fails she will have to write another essay. If a student fails on the first essay and has to write a second, how many different pairs of essay topics can she choose (a) if she is allowed to repeat the same topic? (b) if she is not allowed to write the second essay on the same topic as the first?

Answer
The principle applies here because the student has to choose a first and a second essay topic. In (a) there are six possibilities at each stage, giving a total of $6 \times 6 = 36$. In (b), there are six choices for the first essay, but only five for the second, since the choice has been reduced by one topic. So here the number of possibilities is $6 \times 5 = 30$. Allowing repeats gives more possibilities, a point to which we will be returning later.

9.3 ONE-OR-THE-OTHER PROBLEMS — THE ADDITION PRINCIPLE

Let's return to the problem of choosing a configuration for a personal computing system, and concentrate for the moment on the choice of printer. Suppose that three models are made by one manufacturer, and one model by another. If we do not have any preference for a particular manufacturer, then we have $3 + 1 = 4$ models to choose from altogether. This is an example of the addition principle, which in general tells us that if we are choosing an item either from a set with n_1 elements, or from a disjoint (non-overlapping) set with n_2 elements, or $\ldots$, then the total number of possible choices is $n_1 + n_2 + \ldots$. We can also think of this as the 'or' rule.

There are obviously a lot fewer possibilities in this case than in that of the previous section, since $n_1 + n_2 + \ldots$ is less than $n_1 \times n_2 \times \ldots$ when the ns are positive integers greater than 1. This agrees with common sense — being allowed to combine items from the different sets gives us more choice than having to choose just one item from the combined sets.

The phrase 'combined sets' should make you think of the operation of set union discussed in Chapter 2; in fact the number of possible choices in this kind of one-or-the-other problem is precisely the number of elements in the union of the separate disjoint sets from which we are choosing.

Two further examples:

Sec. 9.3]   **One-or-the-other problems — the addition principle**   223

Example 1
A final-year student must choose one optional subject from three offered by the Accounting department, two offered by Computer Science and two by Economics. How many choices has she altogether?

Answer
Since the student can choose either one from the three accounting topics, or one of the two computer science topics, or one of the two in economics, she has altogether $3 + 2 + 2 = 7$ to choose from.

Example 2
We need to select from the positive integers less than 100 one which is neither even nor a multiple of 5 (this is a selection sometimes used in setting a seed for random number generation by computer. How many possible numbers are there to choose from?

Answer
It is easier here (and in many other cases, as we'll see later) to count the numbers which *can't* be chosen. There are 49 even numbers less than 100, and 10 odd multiples of 5 — the even ones have already been included. We want numbers which are neither in the first group nor in the second. We can use the addition principle to tell us that there are $49 + 10$ choices of a number which won't do. However, as there are 99 integers less than 100 altogether, that leaves $99 - 59 = 40$ to choose from.

Example 2 makes it clear that the addition principle implies a 'subtraction principle', in the following sense. For a well-defined process of choice in which every item under construction is either acceptable or not, we must have

$$\begin{array}{c}\text{number of acceptable}\\\text{choices}\end{array} + \begin{array}{c}\text{number of unacceptable}\\\text{choices}\end{array} = \begin{array}{c}\text{total number}\\\text{of items}\end{array}$$

Thus number of acceptable choices = total number of items − number of unacceptable choices, which is the idea we have used in Example 2.

Although the distinction between the multiplication and addition principles may seem quite clear when they are set out as in the last two sections, it is not always obvious which one should be used in a given situation. So before we go any further, here is a mixed collection of examples for you to try.

Exercise 9.3.1
The writing of a large piece of software is to be divided among a team of eight programmers. When each section of code has been written by one of the programmers, it will be passed on to a different programmer for checking. How many different combinations of programmer and checker are there for any one section of code?

Exercise 9.3.2
The team of programmers in Exercise 9.3.1 consists of five men and three women. How many ways are there of selecting a programmer/checker in which both are of the same sex?

Exercise 9.3.3
How many ways are there of selecting two children from a group of three boys and two girls (a) if the two must be of the same sex? (b) if the choice may be made freely?
(Hint: this is NOT the same situation as in the last two questions — try listing all the possibilities, using letters to denote the various children, and you will probably realize why.)

Exercise 9.3.4
A computer password is to be made up of two letters followed by three digits. To avoid confusion the letters I and O are not to be used. How many possible codes are there (a) if letters and digits may not be repeated? (b) if repeats of both letters and digits are permitted?

Exercise 9.3.5
A piece of electronic equipment contains four circuit boards; it is known that there are two faults somewhere in the equipment, but not where they occur. If a circuit board can contain more than one fault, how many possible distributions of the faults are there? (You may have to answer this question by enumeration.)

9.4 SELECTION METHODS — NOTATION AND TERMINOLOGY

In the last set of exercises you were introduced to several different ways of making a selection, and you have probably begun to discover how these affect the way in which the number of possibilities is counted. We now formally define those different selection methods.

First we have the question of whether or not the order in which items are selected is important. In Exercise 9.3.1, if the eight programmers are denoted by the letters A to H, then the case where A wrote the code and B checked it is clearly not the same as the case where B wrote it and A checked it. If we decided to make the selection by pulling the letters from a hat, adopting the convention that the first name out is the writer, then the order in which the names emerged from the hat would be significant.

The same applies to the construction of computer passwords: ZX425 and XZ245 are clearly distinct, and could be given to two different users without danger of confusion on their part or that of the computer.

In Exercise 9.3.3, on the other hand, we are simply concerned to finish up with a set of two children at the end of the selection process; if the two selected are Mary and Daisy, it does not matter whether Mary or Daisy was chosen first — only the final set is significant.

In fact, the use of the word 'set' here gives you a clue that what we are really discussing is nothing other than the distinction between sets and n-tuples made in Chapter 2. If, in a problem, it is the set of items selected which is of importance, we have an 'order doesn't matter' situation. If, on the other hand, the n-tuple of items $[x_1, x_2, x_3, \ldots]$ constitutes a different solution to our problem from the n-tuple $[x_2, x_1, x_3, \ldots]$, then 'order matters'. There will always be more possibilities in an 'order matters' problem than in an 'order doesn't matter' problem involving the same numbers; for example, if we make an 'order matters' selection of two letters from P, Q, R, S, T, then PQ and QP are regarded as different selections, but if 'order doesn't matter', then these two only yield one selection. To put it another way, one set $[x_1, x_2, \ldots]$ can be re-arranged to provide a number of different n-tuples — just how many re-arrangements are possible is a point we will return to later.

The second question we need to answer is whether or not repeated selections of the same item are allowed. In Exercises 9.3.1 to 9.3.3, this was not the case — the code had to be checked by a different person, and we (implicitly) wanted to select two different children — not the same child twice over. But in Exercise 9.3.4, we first considered the case where repeats were not allowed — so that passwords of the form AA444 would not be permitted — and then allowed such passwords. You should be able to see from your answers to that question that, not surprisingly, allowing repeats

Sec. 9.4]   Selection methods — notation and terminology   225

gives us more possibilities to choose from, and therefore a larger number of selections in total. The same applies in Exercise 9.3.5, where more than one fault may occur in a single circuit board.

These two ways of classifying problems — order matters/doesn't matter, and repeats allowed/not allowed — interact to give four possible cases altogether, as shown in the 2×2 table (Table 9.1).

Table 9.1

	Order matters	Order doesn't matter
No repeats	I	II
Repeats	III	IV

Thus far in our discussion we have mainly looked at problems involving specific numbers, but it is convenient to have a general terminology for referring to selection problems, and some fairly standard conventions have grown up about this. We use n to represent the number of items from which we are making a selection, and r to denote the number of items in the group we are selecting. At first sight it seems that r must be smaller than n, but a bit of thought will reveal that this need not be so if repeats are allowed — we can make 100 selections from the two digits 0 and 1, (that is, $n = 2$, $r = 100$) as long as we are allowed to repeatedly select the digits thus: 1000011101....

We conclude this section by placing each of the exercises from the last section in the appropriate box in Table 9.1, and identifying the value of n and r.

In Exercise 9.3.1, $n = 8$, $r = 2$ (one writer, one checker). No repeats, order matters, so this one goes in box I.

Exercise 9.3.2 belongs in the same box, but $n = 5$, $r = 2$ for men, and $n = 3$, $r = 2$ for women.

Exercise 9.3.3 — we've already seen that order doesn't matter here, but repeats can't be allowed, so the problem goes in box II. $r = 2$ and $n = 5$ for the free choice, 3 for boys, 2 for girls.

In Exercise 9.3.4, $n = 24$ for the letters (no I or O) and 10 for the digits; $r = 2$ for numbers, 3 for digits. The (a) version would belong in box I, the (b) version in box III.

Finally, in the case of Exercise 9.3.5, repeats are allowed (more than one fault can occur in the same board) and there is no concept of order — the faults are identical. So this one goes in box IV, with $n = 4$ and $r = 2$.

Test your understanding of the material of this section by classifying a few problems for yourself — decide where they belong in Table 9.1, and identify n and r, but don't try to solve them yet.

Exercise 9.4.1
A student has to make a choice of three A-level subjects from 10 which are on offer at her school.

Exercise 9.4.2
It is known that some combination of two circuit boards in an assembly of 200 leads to a fault. An engineer has no way of identifying the two boards involved, so he wishes to know how many different pairs of boards he would need to test in order to cover all the possibilities.

Exercise 9.4.3
A consultant producing a report wants to use a different colour paper for the main report, an appendix, and a listing of a computer program. He has green, yellow, blue, pink and white paper available, and wants to know how many different choices are possible.

9.5 FORMULAE AND JARGON

By now you have already calculated, alone and unaided, quite a few results in combinatorics. However, you are probably wondering if it is really necessary to think through every problem form scratch; and while it is not always a good idea to be too attached to formulae without thinking about the theory behind them, it is a healthy instinct to want to generalize a method and establish a principle. So we will now formalize the calculations which have been going on, and establish some general results for short-cutting the counting process.

We follow the order in Table 9.1, and consider first the case where repeats are not allowed and order matters. This was exemplified by the problem (9.3.1) of choosing two people from eight to write/check a piece of computer code. There are eight ways to choose the first person, but only seven ways to choose the second, since repeats are not allowed. We are combining the two in an 'and' manner, so the number of possible combinations is 8×7. Had we needed a third person, the number would have been $8 \times 7 \times 6$, and in general we would have as many factors in the product as there are people being selected. Thus for the case where r things are being selected from n, the result will be $n \times (n-1) \times (n-2) \times \ldots$ possibilities, with r factors altogether, the last of which will be $n-r+1$ (if you can't see why this is so, try checking it by substituting some numbers for n and r).

This slightly clumsy form of expression can be simplified by introducing a piece of symbolism which may be new to you, though it received a brief mention in Chapter 4. We write $n!$ (read 'n factorial') to denote $n \times (n-1) \times (n-2) \times \ldots \times 3 \times 2 \times 1$. $6!$ is thus $6 \times 5 \times 4 \times 3 \times 2 \times 1 = 720$; $3! = 3 \times 2 \times 1 = 6$, and so on. For consistency we also need to define $0! = 1$. (If you have a calculator with a ! function key you can have fun seeing how rapidly $n!$ becomes a very large number as n gets bigger, and how slow your calculator is to work out, say, $20!$.)

At first sight it does not look as if this will help much — our product only goes down to $n-r+1$, not all the way to 1. However, to get the required product starting with $n!$ we need to knock off the unwanted terms $(n-r) \times (n-r-1) \times \ldots \times 3 \times 2 \times 1$, which can also be conveniently written in factorial notation as $(n-r)!$. Thus we can now state that the number of ways of choosing r items from n when repeats are not allowed and order matters is $n!/(n-r)!$

This kind of selection — order matters, no repeats — is called a *permutation*, and you will sometimes see the shorthand ^nP_r or $_nP_r$ used to represent it. So we have shown that

$$_nP_r = \frac{n!}{(n-r)!}$$

The next case in the table is 'order doesn't matter, no repeats', as exemplified by the problem of selecting two children from a group of five. The difference between this and the permutation case is that the two selections D, M and M, D are regarded as identical, whereas they would constitute two different permutations. And this gives us a clue as to how to calculate the number of possibilities in this type of case. If we wanted permutations of two children from five, the result above tells us that the answer would be $5!/2! = 20$. However, in terms of non-ordered selections, re-arrangements of the two selected children among themselves are indistinguishable — and we have just seen that each selection of two could be re-arranged in two different ways. There are therefore only half as many non-ordered selections as there are permutations — or, looking at it another way, every non-ordered selection gives rise to two permutations. So the number of non-ordered selections is only $20/2 = 10$.

We can generalize this argument to see that if we want a non-ordered selection of r things from n, we need to divide $_nP_r$ by the number of ways the r things can be re-arranged among themselves — each non-ordered selection of r items gives rise to as many permutations as there are ways of re-arranging the r things. But the number of rearrangements is easily calculated: there are r ways of choosing an item to be the first, then $r - 1$ ways of choosing the second, and so on down to two ways of choosing the penultimate item, and no choice at all about the last item — only one way it can be chosen. These combined according to the 'and' rule give $r \times (r - 1) \times \ldots \times 2 \times 1 = r!$ ways of rearrangeing r items among themselves. (You can also arrive at this result by using the permutation formula to find the number of ways of choosing r things from r.)

Thus to find the number of non-ordered selections, or *combinations* as they are called, of r things from n, we take $_nP_r$ and divide by $r!$, to get $n!/r!(n - r)!$ — usually written in an obvious notation as $_nC_r$.

Moving on now to the third box of Table 9.1, we come to the case where repeats are allowed, and order matters — the situation with the construction of a computer password. If we consider just the choice of three letters from 24, when repeats are allowed, we can see that there are 24 ways of choosing the first letter, and 24 ways of choosing each of the second and third also, thus $24 \times 24 \times 24 = 24^3$ ways in all. More generally, with n choices for every item from the first to the rth, the number of selections would be $n \times n \times \ldots \times n$ (with r factors) $= n^r$ altogether.

The last box of the table is the trickiest to fill, and here we can exploit a useful relationship. Recall the problem of two faults in an assembly of four circuit boards, with the possibility that both are in the same board. Rather than regarding this as a selection problem — choosing a board or boards to receive the faults — we can turn it on its head and look at it as a distribution problem — we are 'giving out', as it were, the two faults to some of the four boards. The situation will become clearer if you look at Fig. 9.3, which shows the boards as 'pigeonholes', into which the faults, shown as zeros, must be put. The figure, of course, shows only one of the possible arrangements.

As the diagram suggests, only three partitions (which we symbolize as 1s) are needed in order to construct four pigeonholes. The number of possible allocations of the two faults (zeros) among the pigeonholes can therefore be viewed simply as the number of ways of rearranging the three 1s and two 0s among themselves. There are seven things altogether, so if they were all distinct we could rearrange them in 7! different ways (from the argument used earlier in filling box II). However, they are

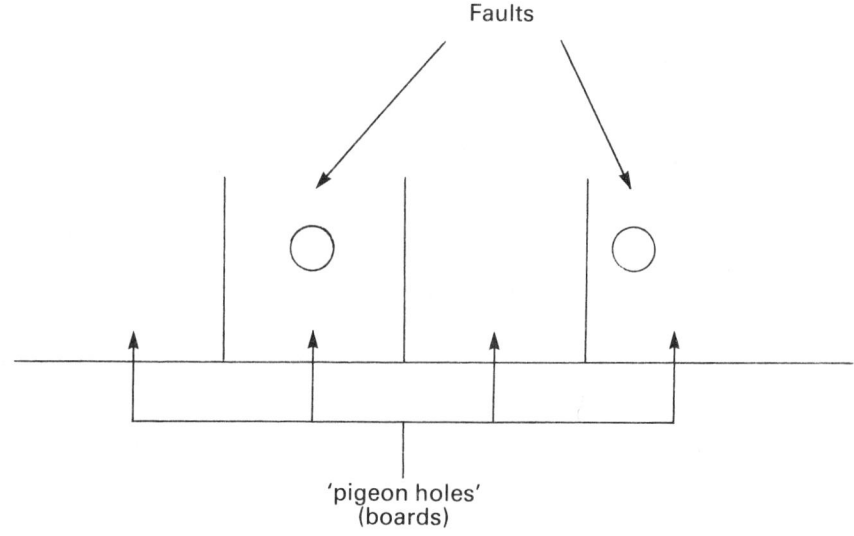

Fig. 9.3.

not all distinct — the three 1s can't be distinguished from one another, neither can the two 0s. So the 3! ways of arranging the 1s among themselves all give rise to the same solution, as do the 2! ways of arranging the 0s. If we allow for this then the number of possibilities reduces to 7!/3!2!

How does this extend to the 'r from n' case? To construct n pigeonholes needs $n-1$ partitions, and these can be arranged among themselves in $(n-1)!$ ways; likewise the r objects being distributed rearrange in $r!$ ways. The total number of items being arranged — partitions plus items — is $n-1+r$, which looks better as $n+r-1$, and so a repeat of the argument we used above, replacing the numbers by letters, gives the number of possibilities as $(n+r-1)!/(n-1)!r!$

This correspondence between selections and distributions in fact extends to other selection methods, though we did not need to make use of it in considering them earlier. Where in a selection problem we have 'repeats allowed', for a distribution we have the possibility of putting more than one item into the same pigeonhole. 'Order matters' translates into the fact that the r items being distributed are indistinguishable. The mathematical term for this kind of correspondence, where the underlying structure of problems, and hence of their solutions, is identical, is isomorphism (you have already met the term in Chapter 5, with reference to graphs). We say that the selection problem and the corresponding distribution problem are isomorphic to one another; recognition of this fact halves the amount of work involved in tackling the two types of problem.

To sum up this section, Table 9.2 is a repeat of Table 9.1, but it now includes the relevant formula in each of the boxes.

We cannot emphasize too strongly the importance of grasping and remembering the underlying theory of each type of selection/distribution, rather than relying on

Table 9.2

	Order matters	Order doesn't matter
No repeats	$n!/(n-r)!$	$n!/r!(n-r)!$
Repeats	n^r	$(n+r-1)!/(n-1)!r!$

'knowing the right formula'. It is far too easy to get confused! Bear this in mind when attempting the following problems, not all of which fit neatly into one of the four boxes of Table 9.2.

Exercise 9.5.1
How many rows has the truth table for $(A \wedge B) \vee (C \wedge D)$?

Exercise 9.5.2
How many subsets has a set with four elements got? Remember that the set itself, and the empty set, are allowable subsets. Why are the answers to this and the previous problem the same? (These two problems are isomorphic.)

Exercise 9.5.3
A class contains five students all named Jones. In how many ways can five assignments labelled 'Jones' be returned to the students so that none receives his or her own work?

Exercise 9.5.4
In how many ways can the letters of the word 'computers' be rearranged among themselves?

Exercise 9.5.5
In how many ways can the letters of the word 'processor' be arranged among themselves? How does this differ from the last problem?

Exercise 9.5.6
Ten people arrive simultaneously to use three checkouts in a supermarket. Regarding the people purely as 'bodies', and not as named individuals, how many ways are there of forming queues at the checkouts? (The number of possibilities will, of course, include some which make no sense from a practical viewpoint, such as all ten people at one checkout and the other two empty.)

9.6 COUNTING AND PROBABILITY

We already saw, at the beginning of the chapter, some of the reasons why being able to count efficiently is important. In the last two sections of the chapter we look at a major application of counting ideas — the calculation of chances or probabilities.

The connection between ways of counting and probability becomes clear as soon as we begin to think about some familiar probability questions, such as the chance of achieving a particular score — say, a total greater than four — in a single throw of a fair die. You would no doubt say immediately that the chance (another word for probability) of such a score arising is 2/6, since there are two scores greater than four out of a total of six possible results altogether. In other words, we count the number of ways that a score greater than four can occur, and express this as a proportion of the total number of results which could arise. We have not, of course, actually defined yet what we mean by probability, but, as we will see, this idea of counting the number of ways a result can arise is fundamental to at least one possible definition.

Before we try to put things on a more formal basis, however, you may want to know why probability is important in the context of computing. After all, the example we just considered suggests that the topic has more to do with gambling than anything else! And indeed it is often easier, in the early stages of studying the topic, to deal with simple problems such as dice-rolling and coin-tossing, rather than with the more complex problems which arise from real-life applications of the theory.

Nevertheless, those applications certainly exist, in computing as in many other areas. We introduce briefly two applications to which we will be returning in more detail later.

First, when businesses are dependent for their successful operation upon access to a central computing facility, it is vital to be able to assess the chances of a link or links in the computer network breaking down. For example, when you use a Cashpoint machine to make an enquiry about your current bank balance, you are using such a link, and it would be very annoying to you if it were not available. Sometimes it may be not merely annoying, but actually dangerous, if such a breakdown occurs, as is the case with the computer networks upon which, in our present unhappy situation, the strategic defence systems of the major nations depend.

Probability theory can enable the chances of various types of breakdown to be assessed, and can also help us decide how much redundancy — extra links to provide a 'safety net' — needs to be incorporated into the system to achieve an adequate measure of protection.

Secondly, consider the case of magnetically coded data which is being read from a computer tape or disc. Because the equipment which writes and reads the data is not completely infallible, and because the tapes or discs themselves are not immune to damage and wear, there is a chance that errors may arise — that a piece of coded information may not be read as representing the character which it was intended to represent. For instance, a 0 might be read as a 1, or vice versa.

Again we need to be able to assess the likelihood of errors of this kind arising, and to devise ways of insuring against the problems they would create. Chapter 10 is devoted to this topic of codes and error-checking, but probability theory provides an important building block in the process, as we will see shortly.

Returning now to the question of what precisely we mean by a probability, we will look in more detail at the problem of scoring more than four with a single throw of a fair die, and see if we can be a little more precise about what's involved. First of all, the throw of the die constitutes a trial (sometimes called an experiment, though the concept is much more general than what is usually conveyed by that term). A trial is a process which can be expected to produce just one of a set of mutually exclusive, exhaustive outcomes. By mutually exclusive, we mean they cannot occur simultaneously; by exhaustive, we mean that between them they cover all the possibilities.

If we apply these ideas to the throw of the die, we see that the set of outcomes is the set of all possible numbers of points showing on the uppermost face of the die when it lands — that is, the numbers from one to six. In the case where we want to achieve a score of more than four, we are interested in only two of these six outcomes, namely those in which five and six points show. Such a set of outcomes is called an event, and denoted by E. If we let S denote the set of all possible outcomes, often referred to as the sample space, then E is a subset of S.

There is one further, rather special feature of the die problem which needs to be mentioned. When we posed the problem, we insisted that the die must be 'fair'. What does this mean? You would probably reply that it means all faces of the die are equally likely to come up on any single throw — that the 'chance' of getting any one score is the same as for all other scores. But 'chance' is just another name for 'probability', so what you would really be saying is that there is a 'something' — a measure of chance or uncertainty — associated with each outcome in the sample space. In the case of the fair die, these measures all happen to be the same, but that will not be the case for all sets of outcomes, as we shall see.

This discussion has clarified what we mean by probability, but has not got us much further towards actually calculating the 'measure of uncertainty' we spoke of. We now need to think about the numerical form such a measure might take.

In the case of the die, of course, you are well aware that the chance of getting, say, a six (or any other score) would be one in six, or 1/6, and the way that figure is reached sheds light on a general method for defining probabilities, at least in some cases. The probability of 1/6 is the ratio of the number of 'successful' outcomes ('successful' simply meaning the ones we are currently interested in) to the total number of outcomes in the sample space. If we denote the sample space by S, and the event (set of outcomes) in which we are interested by E, then in set-theory notation we are saying that the probability of an event, usually abbreviated as $p(\text{event})$, equals $|E|/|S|$. *But* we can only use this definition when, as with the die, all outcomes are equally likely.

What, then, should we do if this is not the case? Or, to put it another way, how should we regard the 'probability' expressed by a computer engineer who tells us that the chance of any one bit being in error when reading data from a magnetic disc is 1 in 1000? How, for that matter, should we react if someone asks us, 'What is the probability that you will live to be 90?' Neither of these probabilities could be worked out along the lines we used for the die; indeed, in the case of your life expectancy, we might say that the set of possible outcomes is infinite (all the different ages on a continuous age-spectrum at which you are might die), and all these outcomes could certainly not be said to be 'equally likely'.

Arguments as to what such probabilities may mean, and what relation they bear to the die-throwing type of probability, have raged for many years among probability theorists, but we will assume that you are more interested in arriving at sensible assessments of the probabilities than in the philosophical underpinnings of the subject. In the case of the probability of a bit being in error when a magnetic disc is being read, the computer engineer has probably (!) used what is called an empirical (in other words, experimental) approach to the problem, reading many thousands of bits of data from disc, counting the number of errors encountered, and estimating the probability in general as number of errors/total number of bits read. Of course, he would have to ensure that the data was read under the same sort of conditions as the user is going to apply, and getting a reliable figure can be very difficult in practice; but leaving aside these problems, the most important point is that he must look at a lot of data. An experimental approach can give us a good approximation to the probability of an event only if the experiment is sufficiently large. (Would you decide a coin was definitely biased if it gave you nine heads in ten throws?)

The problem of estimating life expectancy is even more tricky. A possible 'experimental approach' in this situation is to discover how long a large group of people live, taking into account such factors as their occupation, lifestyle, and so on. This kind of data is collected and tabulated for the use of insurance companies, and you could use it to find out what proportion of people with the same kind of characteristics as you live to be 90, and hence to assess the required probability. But you are not a 'typical' person — you are an individual, and at the end of the day you may prefer simply to give a subjective estimate of the probability, based on your more or less informed judgement. This is the way in which many probabilities are assessed, particularly in the 'soft' sciences, such as management.

It is time to pause and take stock of what we have established so far. If a trial has a set of equally likely outcomes, then

$$p(\text{event}) = \frac{\text{number of outcomes in event}}{\text{number of outcomes in sample space}}$$

$$= \frac{\text{number of ways of succeeding}}{\text{total number of possible results}},$$

to put it more colloquially.

If we cannot use this definition, then we can get an approximation to the probability of an event by doing an experiment (a set of repeated trials), and saying

$$p(\text{event}) = \frac{\text{number of times event is seen}}{\text{total number of trials}}.$$

Finally, if this also fails, then we can rely on a subjective or personal assessment of the probability of the event, using informed judgement.

Here are a few examples to illustrate these ideas.

Example 1
What is the probability that, if a computer engineer knows that there are faults in two boards out of an assembly with 10 boards altogether, but not which boards they are, he will pick the two with faults first time?

Answer
The sample space here consists of all possible choices of two boards from 10. This is a 'no repeats, order doesn't matter' selection problem, such as we examined earlier in the chapter — a combinations problem, in fact, with $n = 10$ and $r = 2$. So the number of possible choices is $_nC_r = 10!/8!2! = 45$. Of these, there is just one which consists of the two faulty boards, so p(engineer picks two faulty ones first time) $= 1/45$. (Notice here that we are using the first definition of probability, which implies an assumption that all choices are equally likely. That might not be the case if, for instance, the engineer is inclined to take the more easily accessible boards first.)

Example 2
You know that in boxes of cut-price floppy discs which you buy regularly, there tends to be a

certain proportion of poor-quality discs. (a) How could you assess the probability that a disc selected at random will be poor quality? (b) In a particular box of eight discs, all have lost their labels, but you know that program A is on one of the discs, and program B on another. What is the probability that if you take out two discs at random, they will be the ones with programs A and B *in that order*?

Answer
(a) The probability of an individual disc being poor-quality could be assessed by collecting data about the proportion of poor-quality discs over a large number of boxes — this would amount to an 'experiment' to determine the probability.
(b) Choosing two discs in a specific order is a 'no repeats, order matters' selection problem — a permutation. With $n = 8$, $r = 2$, the result of Section 9.5 tells us that $_8P_2 = 8!/6! = 56$.

Alternatively, you could say that there are 8 ways of choosing a first disc, then 7 ways of choosing a second, and this is an 'and' situation leading to $8 \times 7 = 56$ possibilities. Of these 56, only one consists of AB, so the probability is 1/56. Again this use of the first definition of probability involves the assumption of a purely random selection of the two discs — every one has the same chance of selection. This takes some organizing in practice, quite apart from being rather unrealistic in this case. Experiments show that if you ask someone to take two items 'at random' from a pile, they tend not to choose items from either end of the pile!

In case (a) above we were interested in finding the probability of just one outcome — a single disc being bad. In (b), however, the event of interest was a combination of the outcomes of two separate experiments — taking a first disc, and taking a second. In this sort of 'compound event' situation, there are two rules for combining probabilities, which can save us a lot of work. Before we look at them, though, let's sum up the characteristics of probability as we have now seen it working.

It is clear that a probability will always be a figure between 0 and 1. Using the first definition, $p(\text{event}) = |E|/|S|$, which will be 0 if the event set is empty — in other words there is no way the event can occur; it is impossible. At the other extreme, if $|E| = |S|$ then $p(E) = 1$. In this case the event set is identical with the sample set, so our 'event' actually consists of all possible outcomes of the trial — one of which is certain to happen. All other possibilities will lie on a scale from 0 to 1 — the closer the probability of an event is to 1, the more probable that event is to occur.

Another useful fact becomes evident if we consider a rather special case. Suppose we are interested in the probability that an event does not occur. We could denote this by $p(E')$, since we are interested in all outcomes except those in E, and set theory tells us that this is precisely E', where the complement is taken with reference to S as universal set. Now $p(E') = |E'|/|S| = |(S - E)|/|S| = 1 - |E|/|S| = 1 - p(E)$.

So we have shown that the probability that an event does not occur is the probability that it does not occur subtracted from 1. This, as we will see, can save effort in tackling problems.

Finally, the sum of the probabilities for all the individual outcomes in the sample space must be 1. For suppose we have n equally likely outcomes in S, then $|S| = n$, and $p(\text{outcome}) = 1/n$ for any one outcome. Thus when n of them are added together they total to 1. The same will apply, by an extension of this argument which you can easily construct, to the probabilities of any set of events which are non-overlapping and between them cover the whole of S. Such events, as we saw earlier, are called mutually exclusive exhaustive events.

9.7 COMBINING PROBABILITIES

Suppose now that we want to find the probability that the score on a die is either even or a three. Here we are combining two separate events — an even score and a three — by means of an 'or' connective. As the diagram in Fig. 9.4 suggests, we can find the

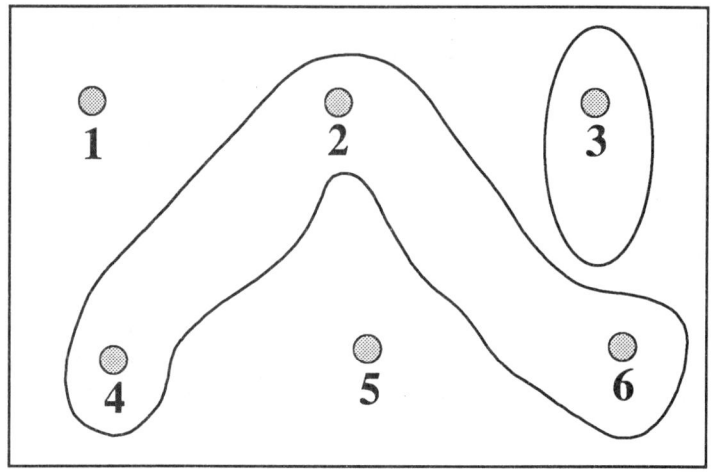

Fig. 9.4.

probability of the combined event by adding together the two separate probabilities $p(\text{even})$ and $p(3)$; thus $p(\text{even or three}) = p(\text{even}) + p(\text{three}) = 1/2 + 1/6 = 2/3$. In general, $p(E_1, \text{ or } E_2) = p(E_1) + p(E_2)$, where E_1 and E_2 are any two mutually exclusive events.

Clearly this would not work if the events were not mutually exclusive — for example, $p(\text{even or a multiple of three})$ can't be found by taking $p(\text{even}) + p(\text{multiple of three})$, since 6 lies in both events and would be counted twice (Fig. 9.5). We have to subtract from the total the probability of the outcome which has been double-counted; here we could say $p(\text{even or multiple of three}) = p(\text{even}) + p(\text{multiple of three}) - p(\text{even and multiple of three})$. The event 'even and multiple of three' is the intersection of the events 'even' and 'multiple of three', and contains just one outcome, namely 6. Thus we have $p(\text{even or multiple of three}) = 1/2 + 1/3 - 1/6 = 2/3$. The 'or' here is thus being used in an inclusive sense — 'one or the other or both' — just as it was in the logical expressions of Chapter 3. There is also a clear connection with the operation of set union; the event E_1 or E_2 is just $E_1 \cup E_2$.

The other way in which we can combine separate events is to insist that two (or more) events occur simultaneously — 'E_1 and E_2'.

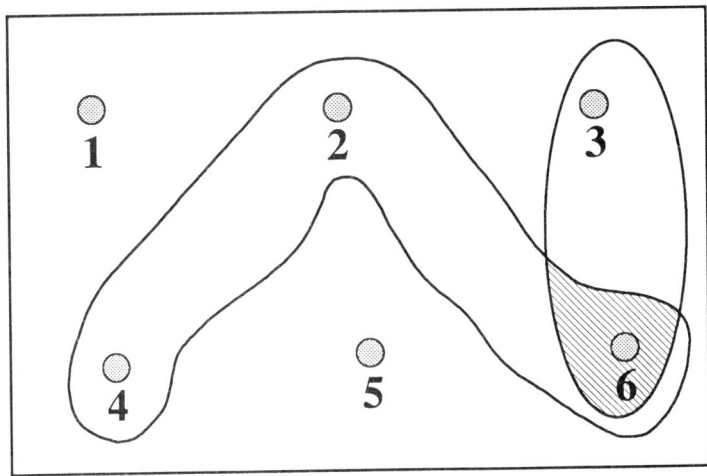

Fig. 9.5.

This, as you might expect, corresponds to logical 'and', and is related to the operation of set intersection; the event E_1 and E_2 is $E_1 \cap E_2$. To find the probability of this event, we multiply the two individual probabilities: p(score on die is even and multiple of three) = p(even) $\times$ p(multiple of three) = $1/2 \times 1/3 = 1/6$. We can see that this is correct, since the only number on the die to satisfy both conditions is six.

Again we have to be a bit careful here — if the chance of E_2 happening depends on whether or not E_1 has happened, we would need to modify this to read $p(E_1$ and $E_2) = p(E_1) \times p(E_2$ given that E_1 has happened). The problem we looked at in the last section, of selecting from a box of eight floppy discs the two containing programs A and B in that order, is a case in point. The probability that the first disc selected is the one containing A will be 1/8 (from the first definition of probability). However, we presumably do not put the first disc back into the box before selecting the second. On the second selection, therefore, we are choosing from only seven discs, so the probability of taking the one containing B is 1/7. We might thus say $p(A$ and B obtained in that order) = $p(A$ taken) $\times$ $p(B$ taken given that A has been taken) = $1/8 \times 1/7 = 1/56$ — which of course agrees with what we obtained earlier.

It is much easier to make sense of these laws when they are put into a practical context; moreover, most newcomers to probability theory find that even when the basic principles have been grasped, it is difficult to develop a 'technique' for tackling problems. So we will give a greater than usual number of worked examples to conclude this section.

Example 1
A communications network consists of six centres A–F linked as shown in Fig. 9.6. If every centre sends roughly the same volume of messages to every other centre, which links in the network will have the heaviest traffic? If the probability of BD breaking down on any given day is 1 in 50, and the probability for DF is 1 in 80, what is the probability that on a particular day both lines between B and F will be open?

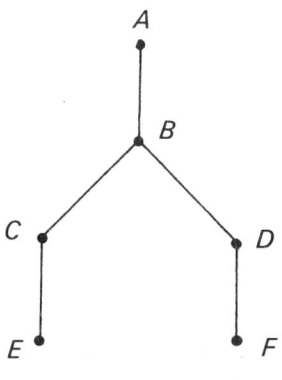

Fig. 9.6.

Answer
The links BC and BD will have the heaviest traffic, since they carry messages going to E and F as well as those for C and D. For both lines between B and F to be open, we want $p(BD$ open and DF open) so we use the multiplication law of probability to get p(both open) = $p(BD$ open) $\times p(DF$ open). Now $p(BD$ broken down) = 1/50, so we can use the fact that $p(E)$ = $1 - p(\overline{E})$ to get $p(BD$ open) = $1 - 1/50 = 49/50$. Similarly $p(DF$ open) = 79/80, and so p(both open) = $79/80 \times 49/50 = 0.96775$, or about a 97% chance.

Example 2
In transmitting an eight-bit binary word, the probability that any one bit is wrongly transmitted is one in 1000. What is the probability that two bits are wrongly transmitted?

Answer
If, say, the first two bits were transmitted wrongly, the required probability would be p(first wrong and second wrong) = p(first wrong) $\times p$(second wrong) = $(1/1000)^2 = 10^{-6}$, using the exponent notation we met in Chapter 4. But it need not necessarily be the first two bits which are in error — it could be bits 5 and 7, or the last two, and so on. In fact we need to multiply the figure of 10^{-3} by the number of combinations of two bits from eight, which the results developed earlier in this chapter tell us is $8!/6!2! = 28$. The probability of two bits in error somewhere in the word is therefore 28×10^{-6}.

The important point about this example is that the probability of n errors diminishes very rapidly as n increases; this is a consequence of the multiplication law of probability. As a result, we can effectively ignore the possibility of getting more than one error — a fact which will be used in Chapter 10.

In both these examples, we have implicitly assumed independence — the chance of line BD going down is not related to the chance of DF going down; the chance of one bit being in error does not change because another bit is in error. Whether these assumptions are actually justified would depend on the circumstances of the particular problem. We now look at an example where independence definitely does not apply.

Example 3
A software house employs 20 analysts and 16 software engineers. Forty per cent of the

analysts, but only 25% of the software engineers, are women. If any one of the staff is equally likely to answer the telephone when you ring the firm, what is the probability that the person who answers is (a) a female analyst (b) a woman?

Answer
In (a), the proportion of women depends on whether the person answering is an engineer or an analyst, so we must say

$$p(\text{female analyst}) = p(\text{analyst}) \times p(\text{female given analyst})$$
$$= 20/36 \times 40/100 = 2/9.$$

In (b), the statement 'person is a woman' can be expanded to 'person is an analyst and a woman or person is an engineer and a woman'. So we have both 'and' and 'or' rules being used, to get

$$p(\text{women}) = p(\text{analyst}) \times p(\text{woman given analyst})$$
$$+ p(\text{engineer}) \times p(\text{woman given engineer})$$
$$= 20/36 \times 40/100 + 16/36 \times 25/100 = 1/3.$$

Finally, a more complex example concerning computer networks.

Example 4
The three sites P, Q and R of a computer network can communicate with each other either directly, or via an intermediate site; thus P can communicate with Q via line PQ or via PR followed by RQ (see Fig. 9.7). The probabilities of breakdown on each line on any given day

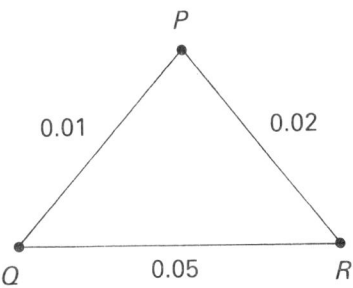

Fig. 9.7.

are shown in the figure. What is the probability that on a given day P and R will be able to communicate? If P and R cannot communicate, what is the probability that line QR is down?

Answer
The statement 'P and R will be able to communicate' looks innocent enough, but it disguises a great variety of possible circumstances, all of which will result in P and R being in communication. Let's begin by listing all possible mutually exclusive combinations of failure/non-failure among the three links. Since any, or all, of the links can fail, the determination of the number of different fail/OK combinations is an 'order doesn't matter, repeats allowed'

problem with $n = 2$ (fail or OK) and $r = 3$ (the number of links). We thus expect to find $2^3 = 8$ different combinations.

These are listed below, using the notation $\overline{PQ}$ to mean that line PQ is out of action; the cases in which P and R can communicate, either directly or indirectly, are marked with an asterisk.

$$\begin{array}{lll} PQ\ PR\ QR^* & \overline{PQ}\ \overline{PR}\ QR \\ PQ\ PR\ \overline{QR}^* & \overline{PQ}\ PR\ \overline{QR}^* \\ PQ\ \overline{PR}\ QR^* & PQ\ \overline{PR}\ \overline{QR} \\ \overline{PQ}\ PR\ QR^* & \overline{PQ}\ \overline{PR}\ \overline{QR} \end{array}$$

So we see there are five different situations in which P and R can communicate. However, before we launch into calculating these five probabilities, let's recall what was said in an earlier section: it sometimes pays to make use of the fact that probabilities total to one, and to find the probability that something does not happen rather than the probability that it does. Here we can look at the probability that P and R cannot communicate, which the list above shows will only involve the calculation of three probabilities — a useful saving in work.

Taking the three separately:

$p(\overline{PQ}\ \overline{PR}\ QR) = 0.01 \times 0.02 \times 0.95 = 0.00019$

$p(PQ\ \overline{PR}\ \overline{QR}) = 0.99 \times 0.02 \times 0.05 = 0.00099$

$p(\overline{PQ}\ \overline{PR}\ \overline{QR}) = 0.01 \times 0.02 \times 0.05 = 0.00001$

As we want these mutually exclusive cases in an 'or' combination we add to get

$p(P$ and R can't communicate$) = 0.00119$.

Thus $p(P$ and R can communicate$) = 1 - 0.00119 = 0.99881$.

You can see that, while all three sites could communicate if only two links (say, PQ and PR) were provided, the inclusion of the extra link provides a significant increase in the reliability of the network. With only PQ and PR, the chance of P being cut off from R is 2%, whereas we have just shown that, even though the extra link QR has a relatively high failure probability of 5%, its inclusion reduces the chance of a PR communication blackout to only a little over 0.1% (not to mention the fact that with only two links, if PR goes down then Q is also cut off from R). For this reason, the extra cost of some links in a network which are strictly speaking redundant is usually worthwhile in terms of improved reliability, though as you can imagine, for a larger network the decisions about which extra links to include can become quite complicated.

The second part of the problem requires us to calculate, given that P and R are unable to communicate, the probability that this is due to QR having failed (either alone, or in combination with another line). Assessing this and related probabilities could assist an engineer in deciding where to start repairs. The easiest way to deal with this question is as follows: the overall probability that P and R are cut off from each other is, as we have just seen, 0.00119. Another way of looking at this is to say that, out of every 100 000 days, there will be 119 when P and R are cut off. (100 000 days is, it is true, a silly figure, amounting to some 274 years, but we use it purely so that the computation can be expressed in whole number terms.)

Out of the three possible situations leading to a communications failure between P and R, which we listed above, two involve failure of QR:

$p(PQ\ \overline{PR}\ \overline{QR}) = 0.00099$

$p(\overline{PQ}\ PR\ \overline{QR}) = 0.00001$.

So $p(QR$ involved in failure$)$ is the sum of these two, 0.001. Thus in 100 000 days, there will be 100 failures of QR.

Now we can use the definition of probability to say that

$p(QR$ down given that P and R can't communicate$) =$

$$= \frac{\text{number of occasions involving } QR}{\text{total number of occasions when } P \text{ and } R \text{ are out of touch}}$$

$= 100/119 = 0.84.$

In other words, there is a good chance that, if P can't communicate with Q, line QR will be down.

We have gone into this example in a good deal of detail partly on account of its obvious practical relevance, but also because it provides good illustrations of the way in which one tackles probability problems in general.

9.8 A CONCLUDING WORD

There has not been time in this chapter to give you more than a first glimpse into the world of probability. Certainly the two laws we have established for combining probabilities, together with the counting methods we looked at earlier, will take you quite a long way, but to deal with more complex problems like those which arise in actual computer networks, more advanced theory is needed. However, what you have learned in this chapter should provide you with a grounding on the basis of which you can go on to read more about the subject if you wish. You will find some books which cover this area listed in 'suggestions for further reading'.

We cannot emphasize too strongly the importance of plenty of practice in order to become at ease with probability ideas. There are not a great many 'methods' to remember; what you need to do is to get into a 'way of thinking' which helps you to decompose problems into their component mutually exclusive events, and to split them into combinations of 'and' and 'or' statements. Remember, too, that there are many other words which can express the sense of 'and' and 'or' — 'neither', for example, really means 'not ... and not ...', while 'both' is another way of saying 'and'. The English language can cause a lot of complications!

PROBLEMS
1. On any day I have the choice of going to work by car, by bicycle or on foot. In a five-day week, how many different combinations of methods of transport are possible?
2. Three software engineers are working on an important project. They must all be present for three consecutive days next week. If the probability of any one of them being absent on a given day is 0.01, and if absences are independent of each other and of the day of the week, what is the probability that all three will be present on one day? What is the probability that the three consecutive days without absence will be achieved?
3. For the problem above, how many ways are there of two person-days being lost in three days?
4. A student in Manhattan lives six blocks south and five blocks west of college.
   (a) How many routes can she take from home to college?
   (b) She always passes a certain corner three blocks north and two blocks east of home. How many different routes are there passing this corner?

5. On a well-known brand of door-key notches may be cut to one of 10 possible depths in any or all of six different positions. How many distinct door-keys can be made?
6. A computer network is shown symbolically in Fig. 9.8; the figures on each link represent the probability that the link is functioning on a given day. What is the probability on a particular day that a message can be sent from A to D? Show that if a duplicate line with the same reliability is put in between C and D, the probability that A and D are able to communicate is improved to over 99%.

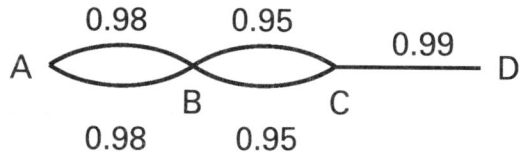

Fig. 9.8.

7. A firm has purchased poor-quality wiring to link its central processor to a remote terminal. As a result the link is only 80% reliable. How many such lines in parallel would be needed to improve this figure to better than 99%?
8. Some types of digital equipment use coding which involves words of length 12 characters, the characters being selected from an 'alphabet' of 256 possibilities. Write down an expression for the number of different words under this system. (Do *not* attempt to work out the answer!)
9. In a system which uses 8-bit binary words to transmit data, the probability of any one bit being received erroneously (i.e. a 0 being received as a 1 or vice versa) is 1 in 10^3.

    (a) If the word 00000000 is transmitted, how many different words involving one error can be received?
    (b) What is the probability that the received word will have one error?
10. A design engineer wishes to test that a new variety of magnetic tape will cope successfully with data written and read on any combination from six different read/write units. How many tests must be run if

    (a) she wishes to test all combinations of read/write unit?
    (b) she wishes to test all combinations except those which involve reading and writing on the same unit.

10
Error-correcting codes

10.1 CODES AND ERRORS

In this chapter we are going to look at the application of some ideas encountered earlier — especially matrix algebra and probability — to the construction of reliable methods for transmitting information in a computerized format. Before we turn to the computer's methods of storing and transmitting data, however, we can begin to get a feel for what's involved by looking at the method we are currently using for transmitting information from ourselves to you — namely, written English.

We assume that the title of this section will have been printed correctly, so that the message you 'receive' is the one which we 'sent'. Had that not been the case, there are several different situations which might have arisen. The title might have been printed as 'Codes and errers' — an error which is easy for you both to detect (if you can spell!) and to correct. There is clearly only one erroneous character in the message, and there is as far as we know only one English word differing from the word 'errers' by a single character. So we simply recognize the message as containing one error, and correct that error by altering the character identified as wrong to give, in some sense, the 'nearest' correct English word.

Now imagine that the section title had been printed as 'Codes and pqrtxz'. In this case, it is easy to see that the message contains errors, since pqrtxz is not a recognizable English word. But it is not easy for you to correct the errors. Even if you assume that the correct number of characters has been printed, there are many six-letter words which would make sense in the context, and you have no way of knowing which of these was intended by us, the senders of the message. So here is a situation where error detection is easy — the word you have received does not match any valid word in the language we are using — but error correction with any degree of certainty is impossible.

Finally, consider the possibility that the title has been printed as 'Nodes and errors'. When you receive that message you have no way even of detecting that an error has occurred, let alone correcting it. The error in printing the word 'Codes' has transformed it into another perfectly valid member of our set of English words, and so it will be accepted at face value as representing the intentions of the sender of the message.

We thus have four possible situations:

(1) A message may be transmitted (in our case, printed) correctly.
(2) It may contain an error which can be both detected and corrected to regain the sender's original message.
(3) It may contain an error which can be dectected but not corrected with any degree of certainty, so that the receiver knows there has been an error but cannot recapture the original message.
(4) It may contain an error which can be neither detected nor corrected, so that the receiver accepts as correct a message which is in fact wrong.

Clearly, each alternative in this list is more undesirable than the previous one, and we should endeavour to ensure if at all possible that only cases (1) and (2) arise.

10.2 SOME IMPORTANT IDEAS

We can extract from the simple example above some important ideas which are applicable also to the transmission of computerized data. First we have the concept of a message composed of words, which are scanned in sequence one by one. In English, these words consist of a variable number of characters (letters) selected from an allowable character set or alphabet. In automated data transmission the characters are also selected from a pre-determined alphabet, but the words are often of a standard fixed length. We will restrict ourselves to this case.

Then we have the notion of a process of transmission and reception of the message word. For simplicity, here and throughout the chapter, we will use 'transmitted word' to indicate the word intended by the sender of the message — the 'correct' word if you like. We will refer to the 'received word' when we mean the word which reaches the receiver; this word may or may not be identical to the transmitted word. If it is *not* identical, then rather than simply saying that 'an error has occurred' irrespective of how many wrong characters the received word contains, we will say that the received word has one error if it differs in a single character from the transmitted word, two errors if it differs in two characters, and so on.

In automated data transmission there are usually two additional stages in the transmission–reception process, called *encoding* and *decoding*. We will be looking in detail at these processes later, but for now you may find it helpful to think in terms of sending a message via Morse code. The message or string of words is encoded into a sequence of dots and dashes, which are then transmitted — usually by electronic means. The receiver has to decode the message, or translate it back from Morse into English, before being able to read it. This is not an exact analogy with the kind of encoding we will be discussing, which fulfils a wider purpose than merely facilitating transmission. However, it should give you an idea of the sequence of processes, shown diagramatically in Fig. 10.1. Notice that it is at the decoding stage that any errors which are detectable and correctable are dealt with.

Finally, we have some conception of a *distance* between two words. We corrected

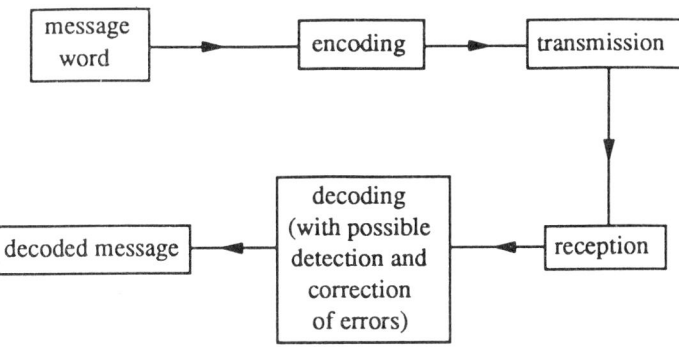

Fig. 10.1.

the word 'errers' to 'errors' rather than to 'apples' on the grounds that a word differing from the received word in only one character is more likely to be correct than one which differs in five characters. We are in this sense, when we correct an error, choosing the *nearest* valid word to the one we have received.

The occurrence of the word 'likely' in the last paragraph suggests that probability may be involved here, and indeed in Chapter 9 you worked an example closely related to the current problem. If the probability of any one character being wrongly transmitted is a very small quantity, which we will call e, and if the errors occur randomly and independently, then the probability of two errors will be of the order of $e \times e$ or e^2. This will be much smaller than e, and the probability of the larger numbers of errors per word will be smaller still. With a six-character word like 'errors', for example, and with $p = 0.01$, you can verify that p (one wrong character in word) $= 6 \times 0.01 \times 0.99^5 = 0.057$, whereas p (two wrong characters) $= {}_6C_2 \times 0.01^2 \times 0.99^4 = 0.0014$. The probability of two errors is about one-fortieth that of one error. And $e = 0.01$ is a high probability of error for a modern data transmission system. So it is reasonable to assume that by far the most likely event is the occurrence of a single error; thus 'errers' is much more likely to have arisen from 'errors' than from 'apples'. (Implicit in this argument is the assumption that the number of characters in the word is not too large; with $e = 0.01$ the chance of getting two errors in a word of length 100 characters would be quite high — but words as long as this are not usual.)

10.3 A SELF-CHECKING CODE

In the example above, you were able to recognize that the received word 'errers' contained an error, because of your knowledge of English. With computerized data transmission, however, this kind of error recognition is not possible. Even in text processing, to program the machine to recognize that a word should have been 'their' rather than 'there' by examining the context would be complicated; and much of the data transmitted by computers is not in the form of recognizable text at all, but consists of strings of alphanumeric characters, in which any sequence appears as likely as any other.

Instead, we need a system in which an algorithm can be used to check whether or not a received word contains an error. Such a system is provided by encoding the message word to be sent. We are going to examine one way of encoding message words, by adding extra characters to them in a manner designed to satisfy certain relationships. If, when the word is received, these relationships are *not* satisfied, then we can deduce that an error has occurred.

The International Standard Book Number (ISBN) system provides a simple illustration of the idea of encoding. These numbers are found printed near the beginning of all recently published books; a book's ISBN identifies the book uniquely. The ISBN is constructed by first allocating to the book a nine-digit number, such as 0-1348-1264, which represents in summary from various facts about the book (for our purposes, the division of the number into three elements can be ignored). To this is added a further character called the check digit, which we will denote by Z, so that the ISBN has the form 0-1348-1264-Z. The choice of a value for Z is made as follows. Form a weighted sum of all the characters of the ISBN including the check digit, using as weights the numbers 10, 9, 8, ..., 2, 1, thus:

$$\begin{aligned}
0 \times 10 &= 0 \\
1 \times 9 &= 9 \\
3 \times 8 &= 24 \\
4 \times 7 &= 28 \\
8 \times 6 &= 48 \\
1 \times 5 &= 5 \\
2 \times 4 &= 8 \\
6 \times 3 &= 18 \\
4 \times 2 &= 8 \\
Z \times 1 &= Z \\
\text{Total} &= 148 + Z.
\end{aligned}$$

Now choose Z so that the total is congruent to zero (mod 11) (if you need to revise the idea of modulo arithmetic, refer back to Chapter 4). Here a value $Z = 6$ will do the trick, since $148 + 6 \pmod{11} \equiv 154 \pmod{11} \equiv 0 \pmod{11}$. So the complete ISBN is 0-1348-1264-6. The original nine-digit number has been encoded into a 10-digit ISBN by the addition of the check digit.

Now let's see what happens if a transposition error is made in writing the number, so that it appears as 0-1438-1264-6. When the weighted total is calculated as before, we have

$$\begin{aligned}
0 \times 10 &= 0 \\
1 \times 9 &= 9 \\
4 \times 8 &= 32 \\
3 \times 7 &= 21 \\
8 \times 6 &= 48 \\
1 \times 5 &= 5 \\
2 \times 4 &= 8 \\
6 \times 3 &= 18 \\
4 \times 2 &= 8 \\
6 \times 1 &= 6 \\
\text{Total} &= 155,
\end{aligned}$$

which is not congruent to zero (mod 11). So the check digit enables us to detect that an error has occurred — though we cannot automatically correct the error. To find exactly where the error has arisen, we would need to look back at the original ISBN and check the copying, to bring the transposition error to light.

It's important to realize that there are two separate processes going on here. One is the *encoding* of a number — the 'message' — by the addition to a check digit; the other is the checking of an ISBN — the 'received word' — for errors. These processes can conveniently be expressed in matrix format; for example, if we write the ISBN as a 1×10 matrix $N = [0\ 1\ 3\ 4\ 8\ 1\ 2\ 6\ 4\ 6]$, and the weights as another 1×10 matrix $H = [10\ 9\ 8\ 7\ 6\ 5\ 4\ 3\ 2\ 1]$, then the matrix product NH' will be a 1×1 matrix — a number — representing the weighted sum of the ISBN digits. Thus for a valid ISBN, $NH' \equiv 0$ (mod 11).

The matrix H which we use to check the validity of the ISBN is called a *parity check matrix*. It gets this name because, in many applications of coding, we add checking characters to a message word in such a way as to satisfy various parity — that is, evenness — relationships; for example, we might ensure that a binary word always contains an even number of 1s. In our ISBN example, however, the 'parity' relation to be satisfied is that the weighted sum of the ISBN digits is zero (mod 11).

We will see in subsequent sections that the idea of encoding a message word by the addition of characters in order to satisfy certain parity relations, and the idea of the parity check matrix, extend to a wide variety of more complex codes.

We call codes like the ISBN system *self-checking codes*, but it should be realized that we have only achieved the self-checking property at a certain cost — the number of valid ISBNs is only one-eleventh of the total set of 10-digit numbers. However, if you care to apply the methods of Chapter 9 to work out the number of valid ISBNs this leaves us, you will see that we are in no danger of running out for some time!

Before we go on to look at other codes, here are a few examples for you to try. One point to note: because the number of possible remainders on division by 11 is 10 — in other words, because we need 11 digits to perform modulo 11 arithmetic — we sometimes need an additional character to represent the check digit 10. The letter X is used in this situation.

Exercise 10.3.1
Find the check-digit for the book numbered 0-8018-1916.

Exercise 10.3.2
Check for errors the ISBNs (a) 0-7458-0280-X; (b) 0-2414-7383-2.

Exercise 10.3.3
Can you devise an error in copying an ISBN which would NOT be detected by the checking procedure? Are such errors likely to be common?

Exercise 10.3.4
If you have a computerized borrowing system at your college or local library, you will find that the number on your borrower's card also contains a check digit. Test yours to see whether the

system is based on modulo 11 arithmetic, and if not, see if you can deduce what the system is — others such as modulo 17 are also in use.

10.4 BINARY CODES

Much of the work done by the world's computers consists of the transmission of data from one place to another. The data is generally transmitted in some kind of coded form; for example, the text which you are now reading was entered into a computer via a keyboard, automatically translated into a string of binary bits via a processor, and then stored on floppy discs. Later it was sent via the processor from floppy disc to printer, to produce the final 'hard copy'.

The coding system which is used in this case is the American Standard Code for Information Interchange (the ASCII code for short), which consists of seven-bit binary words — words of length 7 made up of characters chosen from the set $\{0, 1\}$. There are 128 such words (why?), which are used to represent alphanumeric characters, punctuation and other symbols such as 'new line'. So a short paragraph containing about 800 symbols including spaces would involve some 5600 bits when coded. If during the transmission processes any one bit is for some reason received as a one when it should be a zero, or vice versa, an error will result.

You can see that even if the probability of any one bit being wrongly transmitted is very small, the sheer volume of data involved makes it very unlikely that the received data will be entirely error-free. And while the consequences of an error in transmission of text data such as this may simply be a misprint, in a computer program the result may be a catastrophic failure of the entire program. A 100-line program with about 30 symbols per line will involve about 20 000 binary bits, so again the likelihood of totally correct transmission is virtually nil.

Perhaps it is worth spending a moment looking at the ways in which such errors arise. There are many sources, ranging from electronic noise in transmission lines to reading errors caused by particles of magnetic material clogging the read head of a disc drive. While research is constantly going on to produce more reliable methods of writing, reading and transmitting data, for the foreseeable future technical limitations of time and cost are going to render 100% accurate data handling out of the question. Most systems therefore compromise between the cost of accurate transmission and the cost of error detection and correction, and use some kind of error-correcting code to assist in coping with the inevitable errors.

Not all systems use binary codes, but we are going to limit our discussion to these codes. We need a few further definitions: our codes will have a fixed word-length, denoted by n, which is simply the total number of binary bits in the word. These bits can be sub-divided into two groups: message bits, which are the bits containing the information to be transmitted, and parity bits, which are the additional bits added to assist in error checking. Conventionally we use k to denote the number of message bits, so that the number of parity bits is $n - k$. Thus in the example of the ISBN (not, of course, a binary code), $n = 10$ and $k = 9$, so we have $10 - 9 = 1$ parity bit — the check digit.

Finally, note that we use the term 'the code' to mean the set of all valid codewords.

Exercise 10.4.1
What is the maximum number of binary words available in a code which has $n = 5$?

Exercise 10.4.2
What binary word length would you need to use, as a minimum, if you wish to encode the 26 letters of the alphabet, ten decimal digits and 24 additional symbols?

10.5 THE ERROR MODEL

Before we begin to discuss the construction of error-checking codes, we need to define rather more precisely the kind of error we are considering. In practice it quite often happens that particular faults lead to one more bits in the data being completely unreadable. We will return to this type of error later in the chapter, but for now, we will be much more restrictive, and define an error as occurring when in a binary word a 1 is received as a 0 or vice versa. Thus if the word 0001100 is sent, and is received as 0101100, we say that there is an error in the second bit.

Furthermore, we will continue to assume that the probability of any one bit being in error is very small, so that as explained earlier the possibility that more than one bit in a word is in error may effectively be ignored. This is quite realistic in practice, since error rates as low as 1 error in 10^8 transmitted bits are common today.

We will represent binary words of length n as $1 \times n$ matrices, just as we did for the ISBN earlier. Then, if the word 0001100 is sent, and corrupted in transmission to 0101100, we can write

$$[0\ 1\ 0\ 1\ 1\ 0\ 0] = [0\ 0\ 0\ 1\ 1\ 0\ 0] + [0\ 1\ 0\ 0\ 0\ 0\ 0] ,$$

using the usual definition of matrix addition. In words, this means that

received word = transmitted word + error word ,

where the error word is defined as the difference between the received and transmitted words, and is another word of length n.

Notice that for this definition to be generally applicable, we must operate in modulo 2 arithmetic (NOT the same as binary arithmetic). In this arithmetic, $1 + 1 = 0 \pmod{2}$, so that $0 - 1 = -1 = 1 \pmod{2}$. From now on, all our arithmetic will be done within this system unless we indicate otherwise. Then, for example, if 0111100 is transmitted and 0111000 is received, we can say that

$$\text{error word} = [0\ 1\ 1\ 1\ 0\ 0\ 0] - [0\ 1\ 1\ 1\ 1\ 0\ 0] = [0\ 0\ 0\ 0\ 1\ 0\ 0] .$$

Because of what was said earlier about the virtual impossibility of more than one error in a word, it follows that all error words will contain at most one non-zero bit. So the problem of detecting and correcting an error in a received word reduces to that

Exercise 10.5.1
What is the error word if 1001 is transmitted, and 1011 is received?

Exercise 10.5.2
What is the effect of an error word 00001 on the transmitted word 10101?

10.6 SOME VERY SIMPLE BINARY CODES

Suppose that we want to transmit, in a binary-coded form, messages consisting of the decimal digits from 0 to 9. In binary form, these can be written as

$$\begin{array}{ccccc} 0000 & 0001 & 0010 & 0011 & 0100 \\ 0101 & 0110 & 0111 & 1000 & 1001 \end{array}$$

Notice that we have not yet added any parity bits — we have simply inserted leading zeros to give a fixed word length of four binary bits. Notice also that there are six four-bit binary words which are not included in our list (namely 1010, 1011, 1100, 1101, 1110, 1111).

As things stand, it will not always be possible to detect when errors in transmission occur. Certainly if the word 0010 is sent, but in the transmission process the first bit becomes corrupted so that the word received is 1010, we will recognize this as an error because it is not in our list of valid words. But suppose in sending 0010, it is the second bit which becomes corrupted, so that 0110 is received. Since this *is* a valid word in the above list, the recipient will have no reason to doubt that it is what is sent — in other words, the error will not be detected.

Why is this? It is because the list of valid words contains pairs of words which differ from each other in only one bit, so that a single error is enough to transform one word into another valid word, giving an undetectable error.

We call the number of bits in which two words differ the distance between them. So a pair of words which differ by only a single bit would be said to be a distance 1 apart. If we regard the words as 1×4 matrices, and interpret the arithmetic as modulo 2, then we can define the distance between words x_1 and x_2 as the number of 1s in the 1×4 matrix $x_1 - x_2$. Thus the distance between the words 0001 and 0111 is the number of 1s in 0110, in other words two; but the distance between 0010 and 0110 is the number of 1s in 0100, which as we already saw is one. The number of 1s in a binary word is sometimes referred to as the *weight* of the word, so we can define the distance between two binary words x_1 and x_2 as the weight of $x_1 - x_2$.

It is intuitively clear that the bigger the distance between valid words in a code, the easier it will be to detect when an error has occurred. The idea behind coding is that, beginning with message words some of which may be only a distance 1 apart, we add extra bits to obtain a code in which the distance between valid words is greater than this, making errors more easily detectable — and in some cases correctable. To explore this idea a little, we will consider another very simple code — in fact, about the simplest there is.

One way of trying to increase the chance that a message is correctly received is to send it several times over. So, instead of sending a single bit 1, we might attach a parity bit which is identical with the message bit, and send 11 instead. Likewise a 0 would be sent as 00. This gives us a twofold *repetition code*, with just two codewords 00 and 11. In the original code the distance between the words of the code — 0 and 1 — was 1; in the new version the distance is 2. Whereas in the old code, if a 1 was received instead of a 0, we would accept it as correct and thus make an error, in the new version a single bit error in, say, the word 00 might give us 01 instead. Because the distance between the two codewords is 2, a single bit error *cannot* transform one valid codeword into another, and so such an error will always be detectable.

Such an error is not, however, correctable. If we receive the word 01, we cannot tell whether it is the first bit which is in error — that is, the word sent was 11 — or the second — the word sent was 00. The received word is equidistant between two valid codewords, and there is no means of deciding which is correct.

This idea is made more explicit by the graph shown in Fig. 10.2(a). Here the nodes represent all the possible two-bit words, both valid and invalid, and the edges link pairs of nodes which differ by only a single bit, and are therefore a distance 1 apart. You can see that 01 is linked by a path of length 1 to both 00 and 11, so it is equidistant from the two valid codewords, as already noted.

Now consider what happens if instead of a twofold repetition code we use a threefold one, containing the words 000 and 111. Here a single bit error in 000 could transform it into 001, but the received word is now *nearer* (with the definition of distance introduced earlier) to the correct word than to the other word in the code. We can therefore both detect the error as having occurred, and correct it to its 'nearest neighbour' 000 — thus retrieving the word which was sent. Again, Fig. 10.2(b) interprets this fact by means of a graph.

To recap what we have learned so far: the possibility of detecting and correcting errors in transmission of codes appears to depend on the distance between valid words in the code. If we can increase that distance, by the addition of parity bits to the message bits, we can improve the error detecting and correcting properties of the code.

However, the repetition codes we have looked at are are not very efficient — there is a lot of redundancy in them, in that with the threefold repetition code, for example, we have to send three times as many bits as the actual message contained. In the next section, we look at another more efficient binary code, and deduce some general methods for error detection and correction.

Exercise 10.6.1
How many errors could be (a) detected and (b) corrected, by a four-fold repetition code?

Exercise 10.6.2
Repeat the last question for a five-fold repetition code. Can you see any pattern emerging?
(We will return to the question of the number of errors which can be detected and corrected by a code later in the chapter.)

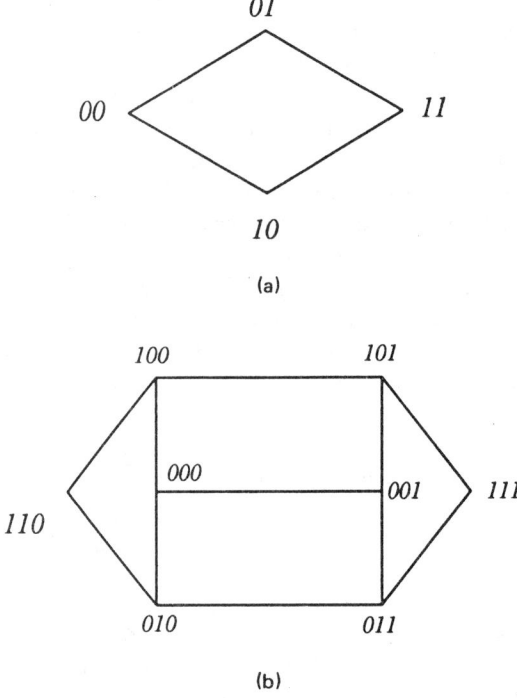

Fig. 10.2.

10.7 A TYPICAL ERROR-CORRECTING BINARY CODE: THE ENCODING PROCESS

Suppose now that the 'words' we wish to transmit consist of all four-digit binary numbers (of which there are 16, as those of you who have read Chapter 9 should be able to verify). Sending these numbers as they stand will lead to the kind of problems described in the last section — namely, some pairs of words are separated only by a distance of one, and so a single-bit error is enough to transform one valid word into another. It is these pairs, separated by distance one, which are of greatest concern, since single-bit errors are much more likely than others. We say that this set of words has *minimum distance* one. As explained earlier, our objective is to increase this minimum distance by the addition of parity bits, in order to improve our chances of detecting and correcting errors.

The problem therefore falls into two parts. First, how can we add the parity bits in a methodical way, in order to achieve our error-correction objective? This is the problem of *encoding* the words we wish to transmit. Second, having transmitted one of the encoded words, how can we reconstruct from the word which is received the one which is transmitted — including, if necessary, the correction of any errors which have occurred in transmission? This is the problem of *decoding* the received message.

We must remember, too, that as the processes of encoding and decoding are to be carried out within the computer, and should be transparent from the point of view of the user, we need efficient algorithms for the encoding and decoding functions.

We will tackle the encoding process first. The immediate question arising is 'How many parity bits should we add?' Intuition suggests that the more parity bits added, the greater we can make the minimum distance of the resulting code, and the better the performance of that code in terms of error-correction properties. However, the corresponding drawback is that if we add, say, six parity bits to each message word, so that the codewords have 10 bits altogether, we will be sending 10 bits to get every four bits of information across. We say that the *rate* of the code is 4/10, and it is clear that in terms of the amount of infornmation transmitted per unit time, the process will not be very efficient. In fact, the nearer the rate is to 1 the better, from this standpoint. There is therefore a trade-off to be made between efficient (high-rate) data transmission and accurate (error-free) data transmission.

We will return to the general issue of the ratio of parity bits to message bits later in the chapter. For the time being, we choose to add three parity bits to our four message bits, to get a code with a seven-bit word length. We will call the message bits x_0, x_1, x_2, x_3, and the parity bits x_4, x_5, x_6, and will write the complete word as $x_0 x_1 x_2 x_3 x_4 x_5 x_6$. (It may seem odd to number the bits from 0 to 6 rather than 1 to 7, but this is the way electronics and communications engineers generally do it. You also need to take care in determining whether the bit-numbers are read from left to right or vice versa. Here we use left-to-right numbering, in line with the notations we have used in studying matrices. Many electronics and communication theory texts, however, number bits from right to left, and while this does not alter the theory it does necessitate changes to the format of some of the results.)

Of course, we are not going to add the parity bits in an arbitrary way — there must be a rule which determines their values, and which can be used to check for the occurrence of errors. For example, in the case of the repetition codes of the last section, the parity rule could be roughly expressed as 'all bits are the same', and more precisely formulated as $x_0 = x_1 = x_2 = \ldots$. One possible set of parity equations for our new code is as follows (three equations are needed to determine the values of three parity bits):

$$x_4 = x_1 + x_2 + x_3$$
$$x_5 = x_0 + x_2 + x_3$$
$$x_6 = x_0 + x_1 + x_3$$

All arithmetic is, as usual, modulo 2. You should verify that this gives rise to the set of sixteen valid codewords listed below:

```
0000000    1000011
0001111    1001100
0010110    1010101
0011001    1011010
0100101    1100110
0101010    1101001
0110011    1110000
0111100    1111111
```

Exercise 10.7.1
What is the minimum distance of this code? Is it necessary to examine all possible pairs of codewords in order to work this out? (There are 16 × 15/2 such pairs, as you should be able to calculate if you've read Chapter 9!)

Of course, working out what the parity bits for each message word should be by using the three separate equations is very inefficient. But note that in matrix notation the parity bits are determined from the message bits by

$$[x_4 \ x_5 \ x_6] = [x_0 \ x_1 \ x_2 \ x_3] \begin{bmatrix} 0 & 1 & 1 \\ 1 & 0 & 1 \\ 1 & 1 & 0 \\ 1 & 1 & 1 \end{bmatrix}$$

or [parity bits] = [message word] × T',

where T is the matrix of coefficients of the parity equations, and T' as usual denotes the transpose of T. So the message word can be encoded in a single matrix multiplication, thus:

$$[x_0 \ x_1 \ x_2 \ x_3 \ x_4 \ x_5 \ x_6] = [x_0 \ x_1 \ x_2 \ x_3] \begin{bmatrix} 1 & 0 & 0 & 0 & 0 & 1 & 1 \\ 0 & 1 & 0 & 0 & 1 & 0 & 1 \\ 0 & 0 & 1 & 0 & 1 & 1 & 0 \\ 0 & 0 & 0 & 1 & 1 & 1 & 1 \end{bmatrix}$$

Exercise 10.7.2
Verify that this matrix multiplication correctly encodes the message word 1001 as 1001100.

We call the matrix in this multiplication the *generator matrix* of the code, and denote it by G. Thus we have the relation

codeword = message word × G.

It is easy to see how G is constructed if we write it as what is called a *partitioned matrix*:

$$G = \begin{bmatrix} 1 & 0 & 0 & 0 & | & 0 & 1 & 1 \\ 0 & 1 & 0 & 0 & | & 1 & 0 & 1 \\ 0 & 0 & 1 & 0 & | & 1 & 1 & 0 \\ 0 & 0 & 0 & 1 & | & 1 & 1 & 1 \end{bmatrix} = [I \mid T'],$$

where I is a 4 × 4 unit matrix.

Sec. 10.8] A typical error-correcting binary code: the decoding process

Partitioned matrices can be multiplied in a particularly simple manner, as an example will illustrate. Suppose we want to perform the matrix multiplication

$$\begin{bmatrix} 1 & 3 \\ 3 & 2 \end{bmatrix} \begin{bmatrix} 1 & 2 & 1 & 0 \\ 0 & 1 & 0 & 1 \end{bmatrix}$$

Multiplied in the normal way this gives $\begin{bmatrix} 1 & 5 & 1 & 3 \\ 3 & 8 & 3 & 2 \end{bmatrix}$, as you should verify. But if we partition the second matrix to get $\begin{bmatrix} 1 & 3 \\ 3 & 2 \end{bmatrix} \begin{bmatrix} 1 & 2 & | & 1 & 0 \\ 0 & 1 & | & 0 & 1 \end{bmatrix} = [A \times B | I]$, say, then we can see that the product is just $[AB|AI]$. In the same way, the process of encoding a message word X will give us $XG = X[I|T'] = [XI|XT']$. The multiplication by I leaves the four message bits unchanged, while the multiplication by T' adds the parity bits according to the defined parity equations.

Notice that G is a 4×7 matrix, where 4 is the number of message bits and 7 is the length of codewords. In general, G will be a $k \times n$ matrix.

Exercise 10.7.3
Write down the generator matrix for encoding the three-bit binary words with the addition of three parity bits which satisfy the relations

$$x_3 = x_0 + x_1 \qquad x_5 = x_1 + x_2$$
$$x_4 = x_0 + x_2$$

and use it to encode the words 010, 110.

10.8 A TYPICAL ERROR-CORRECTING BINARY CODE: THE DECODING PROCESS

We now have a method for encoding any message word X: simply multiply it by G. Now, what about decoding the word — call it Y — which is received after transmission, and which may have acquired errors on the way?

This seems a daunting task, particularly when we realize that there are $2^7 = 128$ possible seven-bit codewords, of which only 16 are valid. To make the problem more concrete, suppose that we wish to transmit the codeword 1001100, which as we found above is the encoding of the message word 1001. However, in the transmission process the sixth bit becomes corrupted so that 1001110 is received. A glance at the list of valid codewords shows that this is not among them — so an error has definitely occurred. But in which bit or bits?

Fortunately we added the parity bits precisely to help us answer this question. For any valid codeword, all three parity relations will be satisfied, so that

$$x_1 + x_2 + x_3 = x_4$$
$$x_0 \quad\ \ + x_2 + x_3 = x_5$$
$$x_0 + x_1 \quad\ \ + x_3 = x_6$$

Since addition and subtraction are equivalent in modulo 2 arithmetic, these relations can be rewritten as

$$x_1 + x_2 + x_3 + x_4 \quad\quad\quad = 0$$
$$x_0 \quad\ \ + x_2 + x_3 \quad\ \ + x_5 \quad\quad = 0$$
$$x_0 + x_1 \quad\ \ + x_3 \quad\quad\quad + x_6 = 0$$

expressing the fact that these sets of bits have *even parity* (that is, sum to zero). In matrix notation this is

$$\begin{bmatrix} 0 & 1 & 1 & 1 & 1 & 0 & 0 \\ 1 & 0 & 1 & 1 & 0 & 1 & 0 \\ 1 & 1 & 0 & 1 & 0 & 0 & 1 \end{bmatrix} Y' = HY' = \underline{0}$$

where Y denotes the received word, and $\underline{0}$ denotes a 3×1 zero matrix.

Now we have come across something like this before, when we were checking ISBNs. The matrix H here is another parity check matrix, having the property that for any valid codeword Y, $HY' = \underline{0}$. It is not unique, so we say it's 'a' parity check matrix rather than 'the' parity check matrix for the code; we could add/subtract the original parity relationships to obtain others, equally valid, which in turn give rise to different parity check matrices. However, for our code with $n = 7$, $k = 4$, all of them will be 3×7 matrices. In general, a parity check matrix for a code with k message bits and $n - k$ parity bits will be an $(n - k) \times n$ matrix.

We can now check our received word Y for correctness by computing HY'; if the product is *not* zero, an error has occurred. (Note that this implication cannot be reversed: the fact that the product HY' is zero does not prove that there has been no error. For example, three errors in the word 0000000 could transform it into 0010110, another valid codeword. In the terminology of Chapter 3, $HY' \neq \underline{0}$ is a sufficient, but not a necessary, condition for an error to have occurred.)

So far, so good. We now have an algorithm for detecting the occurrence of some errors — namely, those which do not convert the transmitted word into another valid codeword. But, faced with the fact that $HY' \neq \underline{0}$ for the value of Y which we have received, we would like not merely to be able to say, 'Oh dear, there's been an error', but to identify which bit or bits contain the error, and correct them to regain the original message word.

To see how this might be done, we need to remind ourselves of the error model introduced earlier in the chapter. We said that received word = transmitted word + error word, so that if we write Z to denote the error word, and X and Y as before for

Sec. 10.8] A typical error-correcting binary code: the decoding process

the transmitted and received words, then $Y = X + Z$. Thus $HY' = H(X + Z)' = 0 + HZ' = HZ'$, since $HX' = 0$ because X was a valid codeword.

We set $HY' = S'$, where S is called the *syndrome* of the erroneous word. The syndrome of a disease, as you probably know, is its set of symptoms, and in a similar way the syndrome of Y contains the symptoms of the error in Y. But the argument above showed that $HY' = HZ'$; so $S' = HZ'$, indicating that S depends only on the error word Z and not on the word which was originally transmitted. This is important because it means that the receiver can compute S completely from information which is available to her.

Example
Find the syndrome of the received word 1001110.

Answer
Using the parity check matrix obtained above, we have

$$\begin{bmatrix} 0 & 1 & 1 & 1 & 1 & 0 & 0 \\ 1 & 0 & 1 & 1 & 0 & 1 & 0 \\ 1 & 1 & 0 & 1 & 0 & 0 & 1 \end{bmatrix} \begin{bmatrix} 1 \\ 0 \\ 0 \\ 1 \\ 1 \\ 1 \\ 0 \end{bmatrix} = \begin{bmatrix} 0 \\ 1 \\ 0 \end{bmatrix},$$

so the error syndrome is $[0 \; 1 \; 0]$.

Exercise 10.8.1
Compute the syndrome of the received word 0101111.

We have seen that S can easily be calculated, while H is a known matrix, so it looks as if we should, in principle, be able to solve the equation $S' = HZ'$ to find Z. Once that is done, we can use the relationship $Y = X + Z$ to determine the transmitted word X from Z and the received word Y.

However, S' contains only three elements, while Z is a 7-bit word, so we have three equations to determine seven unknowns — not very promising. Looked at another way, there are many possible values of Y among the 112 invalid seven-bit words (that is, 128 seven-bit words minus 16 valid codewords) all leading to the same syndrome.

In order to determine Z uniquely, we need to recall another fact about our error model — namely, as argued in section 10.5, that the probability of more than one error per received word is so small as to be negligible. This means that Z will contain at most one non-zero bit; in the terminology we introduced earlier, its weight will be one. To be definite, let's say that $z_i = 1$ and $z_j = 0$ for $j \neq i$.

256   **Error-correcting codes**   [Ch. 10]

Now return to the equation $S' = HZ'$. By the same sort of process as we used above for partitioned matrices, we can rewrite this as

$$S' = H_1 z_1 + H_2 z_2 + \ldots + H_7 z_7 ,$$

where H_j denotes the jth column of H. But we have just said that all the zs are zero except z_i, which is 1, so the equation reduces to $S' = H_i z_i = H_i$.

An example should make this clearer. Suppose in a particular case the error word $Z = 0000100$ had arisen. Then

$$HZ' = \begin{bmatrix} 0 & 1 & 1 & 1 & 1 & 0 & 0 \\ 1 & 0 & 1 & 1 & 0 & 1 & 0 \\ 1 & 1 & 0 & 1 & 0 & 0 & 1 \end{bmatrix} \begin{bmatrix} 0 \\ 0 \\ 0 \\ 0 \\ 1 \\ 0 \\ 0 \end{bmatrix}$$

$$= \begin{bmatrix} 0 \\ 1 \\ 1 \end{bmatrix} \times 0 + \begin{bmatrix} 1 \\ 0 \\ 1 \end{bmatrix} \times 0 + \begin{bmatrix} 1 \\ 1 \\ 0 \end{bmatrix} \times 0 + \begin{bmatrix} 1 \\ 1 \\ 1 \end{bmatrix} \times 0 + \begin{bmatrix} 1 \\ 0 \\ 0 \end{bmatrix} \times 1$$

$$+ \begin{bmatrix} 0 \\ 1 \\ 0 \end{bmatrix} \times 0 + \begin{bmatrix} 0 \\ 0 \\ 1 \end{bmatrix} \times 0 = \begin{bmatrix} 1 \\ 0 \\ 0 \end{bmatrix}$$

= the fifth column of H, since it was the fifth bit of Z which was non-zero.

What we have proved in general, then, is that the syndrome is precisely equal to the ith column of H, *where it is the ith bit of the error word which is non-zero*. So knowing the syndrome enables us, by looking at H, to identify the error word and hence reconstruct the original transmitted word X. Moreover, as a consequence of our assumption that there is only one error per word, we can now say that an error has occurred if and only if $HY' \neq 0$.

Our algorithm for decoding can be summarized thus:

(1) Calculate $S' = HY'$ for the received word Y.
(2) Inspect H to see which column is identical to S'; call this the ith column.
(3) Then Z has a 1 as its ith element, and all its other elements are zero.
(4) But $Y = X + Z$, so $X = Y - Z = Y + Z$ (mod 2 arithmetic, remember!). More concisely, simply change the ith character of Y from a 0 to a 1 or vice versa in order to recover the transmitted word X.

Let's apply this algorithm to our earlier example, for which Y was 1001110 and S was (0 1 0). Since for the code under consideration

$$H = \begin{bmatrix} 0 & 1 & 1 & 1 & 1 & 0 & 0 \\ 1 & 0 & 1 & 1 & 0 & 1 & 0 \\ 1 & 1 & 0 & 1 & 0 & 0 & 1 \end{bmatrix}$$

we see that S is identical with the sixth column of H. So the non-zero bit in Z must be the sixth, giving $Z = 0000010$, and hence $X = 1001100$. If you check back to the beginning of this section, you will find that this was indeed the transmitted word we began with. One question which may have occurred to you is, what if the syndrome is identical with more than one column of H — in which case we would fail to identify uniquely the non-zero element in Z? The only answer we can give to this question, at this introductory level, is that the methods of construction of the binary codes which we are considering, and their parity check matrices, are such that this ambiguity cannot occur.

Exercise 10.8.2
Apply the algorithm we have just derived to test for validity and, where necessary, correct the following received words: (a) 1000011; (b) 0000001; (c) 0001100.

10.9 SOME GENERAL POINTS

We would not wish you to think that the methods we have outlined in the last two sections will work for *any* code, with *any* set of parity equations. The code we chose to look at had many special features, as you probably realized. In this section we will examine some of these features, and discuss how parts of the methods could be generalized to cover other codes.

Codes on larger character sets

Perhaps the most obvious limitation of the code we discussed was that it is a binary code — that is, the codewords are composed of characters selected from the character set $\{0, 1\}$. The method of syndrome decoding can be extended to deal with codes on larger character sets, though many codes used in practice actually are binary.

Linearity

Less obvious, but of great practical importance, is the fact that our example code is *linear*. There are various ways of defining a linear code, some of which rely on an understanding of mathematics beyond the scope of this book, but perhaps the simplest definition is that the parity equations we used were linear equations with constant term equal to zero. A consequence of this is that any *linear combination* $aX_1 + bX_2$ of two valid codewords X_1 and X_2, (a and b being constant numbers) gives a word which is also in the code.

This cannot occur unless the original set of k-bit message words — before the parity bits were added — included all 2^k possibilities. For example, a code with $k = 5$ containing only 26 message words could not be linear, since there are $2^5 = 32$ possible five-bit message word in all.

Linearity has the consequence that the difference between any pair of codewords will itself be a codeword, $X_1 - X_2$ being a special case of a linear combination. So in order to find the minimum distance of the code, it is not necessary to look at every pair of words in the code — the minimum distance will simply be the weight of the lowest-weight codeword(s) (see Exercise 10.7.1).

It can be shown that yet another consequence of linearity is the fact, cited earlier, that all columns of the parity check matrix are distinct. As we saw, this feature is crucial to our decoding algorithm, since it enables us to say 'find *the* column of H which is identical with S', without worrying that there might be two or more such columns.

Systematic property

Our example code is also a *systematic* code, which means that the original message bits appear explicitly as bits in the codeword, and always in the same positions. In our example they were always the first four bits, but this is not essential — they could, for instance, have appeared as bits 0, 2, 3, 5, though the coding and encoding algorithms would require some modifications if this were the case. It can be proved that linear codes are always systematic, though the converse is not true.

Minimum distance and multiple random errors

The example code had minimum distance three, and was therefore able to detect and corect a single error per codeword. This was adequate for decoding under our model, which assumed that the probability of more than one error in a received word was negligibly small.

However, there are situations in which this might not be the case, and where we might wish to detect/correct multiple errors. There is clearly a connection, as we have said before, between minimum distance and the number of errors which can be detected/corrected. We will now look at this connection rather more formally.

You may recall that in Exercises 10.6.1 and 10.6.2 you were asked to think about the number of errors which could be detected and corrected by the four- and five-fold repetition codes. You should have concluded that a five-fold code, which has minimum distance 5, can detect and correct up to two errors per word. This is because with two errors, the nearest neighbour of the received word among the valid codewords is still the transmitted word — other valid codeword will be distant at least three from the received word. Fig. 10.3 shows this situation diagrammatically.

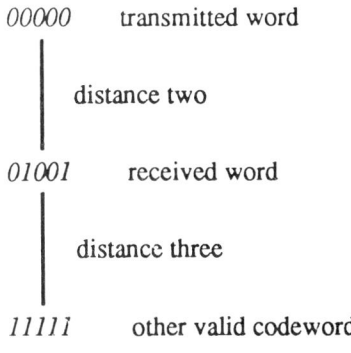

Fig. 10.3.

However, for the fourfold repetition code, two errors can be detected but only one corrected. Two errors result in a received word equidistant between two valid codewords, and it is impossible to use the 'nearest neighbour' argument to decide which of these was sent.

We can generalize this argument to conclude that for a code with minimum distance d, the error-detecting and correcting properties will depend on whether d is odd or even. In the odd case, $(d-1)/2$ errors can be both detected and corrected, but if d is even, $d/2$ errors are detectable but only $(d/2) - 1$ are correctable. (Check that this agrees with what we found for the repetition codes.)

The length of codewords

When we constructed our example code in Section 10.7, we decided to add three parity bits to the four bits of original message words in what looked like a fairly arbitrary manner. Thus the rate of our code — the ratio k/n which indicates how efficient the code is — was 4/7. However, for codes of the kind we've been considering, which can correct a single error per word, it is possible to obtain a relationship between n and k which determines the theoretical maximum value of k — or the minimum number of parity bits required — for a given word length n. We will now demonstrate this relationship.

Suppose we want to encode all the k-bit binary message words, of which we know there are 2^k, into n-bit codewords. There will then be 2^k valid codewords, out of 2^n n-bit words altogether. A valid codeword can be changed into one of its nearest neighbours, at a distance of one, by altering one of its n bits. Thus every valid codewords has n nearest neighbours.

Now, if the code is to be able to detect and correct one error per word, the set for any other codeword X_1 must not intersect with the set for any other codeword X_2 (see Fig. 10.4) — that is, the set of n nearest neighbours of each valid codeword must be

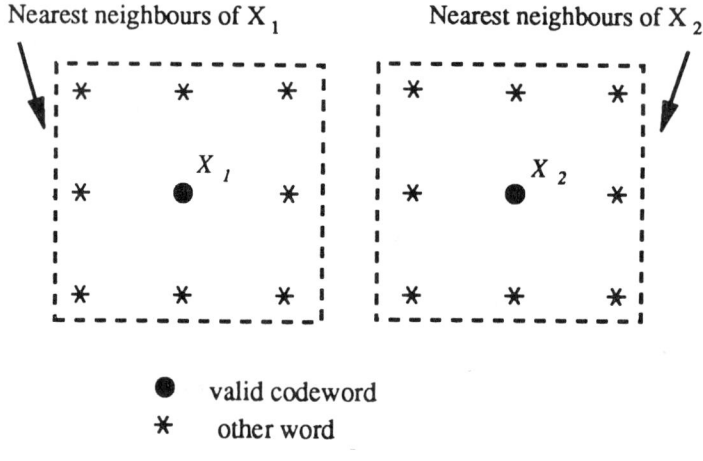

Fig. 10.4.

unique to that codeword. If there were an element common to the sets of nearest neighbours for X_1 and X_2, the 'nearest neighbour rule' would not enable us to correct it without ambiguity, and so single-error correction would not be possible.

But if each of 2^k codewords is to have its own unique set of n nearest neighbours, the set of 2^n n-bit words must contain at least $2^k + n \times 2^k$ words — that is, 2^k valid codewords, plus $n \times 2^k$ nearest neighbours. Thus we can say that

$$2^n \geq 2^k + n \times 2^k ,$$

or more tidily

$$2^n \geq 2^k (1 + n) .$$

This relationship determines a theoretical limit to the rate of an n-bit single-error-correcting binary code. We say 'theoretical' because the result does not provide us with a method for actually producing such codes. This is one of the continuing difficulties with coding theory in general: there is a wide gap between what theory tells us can be achieved in terms of coding and decoding, and practical algorithmic methods for carrying out these processes.

If the result above is re-written as

$$2^n/(1+n) \geq 2^k,$$

then we can compute the theoretical maximum values of k for given values of n, and hence the maximum rates of codes with various word lengths. The table below shows the first few values of n and the corresponding values of k and the rate.

n	3	4	5	6	7
k	1	1	2	3	4
Rate ($=k/n$)	1/3	1/4	2/5	1/2	4/7

Other error modes

The assumption that errors occur randomly and independently formed an essential feature of our error detection and correction method, enabling us to rule out the probability of two or more errors in a word as being very small. Quite often, however, this assumption will be invalid; many errors are of the type known as *burst errors*, in which a whole group of consecutive bits become corrupted. Typically this will happen because of a 'head clog', where particles of magnetic medium become attached to the read/write head of a tape or disc drive.

In this situation we can no longer assume at worst one error per word. But by using an ingenious device, the methods of error detection and correction we examined earlier can still be used to deal with the problem. To see how the device works, suppose for simplicity that the data we wish to transmit consists entirely of zeros, arranged in the form of three seven-bit words, thus:

        0000000   0000000   0000000

If the words were transmitted like this, and a burst error corrupted four bits in the second word, we would have

        0000000   0011110   0000000 ,

and our single-error correcting methods would fail on the second word. So instead of transmitting the words bit by bit in order, the data to be transmitted is arranged in the form of 3×7 array, with the words one above the other;

        0000000
        0000000
        0000000 .

The bits are then transmitted column-wise rather than row-wise — that is, in the form of seven three-bit words rather than three seven-bit words; so when the burst error occurs, giving

```
0000000
0011110
0000000 ,
```

there will only be a single-bit error in each of four of the transmitted three-bit words. These four can therefore be corrected. The receiver can then, having reconstructed the array, obtain the original seven-bit words by reading row-wise.

With this method, a single-error-correcting code could cope with burst errors up to seven bits in length in this example. The method is called *interleaving*, and more complex versions can deal with longer bursts, and with multiple-error correcting codes.

Erasures are another possible occurrence. Whereas in the usual mode of error, a 1 is read for a 0 or vice versa, in an erasure the character becomes totally illegible and can only be regarded as a ? (perhaps because an electrical or magnetic 'blip' is too weak to be read with certainty). Modifications to the kind of codes we have been discussing can also deal with this type of error.

10.10 CONCLUSIONS

In this chapter we have attempted to give you a very brief introduction to the enormous subject of coding theory, using a simple code to illustrate the ideas involved. The code we used, with $k = 4$ and $n = 7$, is an example of a *Hamming code*, a special class of binary, linear, single-error-correcting codes for which error detection and correction can be carried out in the way we outlined. These codes, which are called after their inventor, form a very important group; but there are other codes not of this type for which alternative encoding and decoding algorithms must be sought. Even when such algorithms exist, they are often difficult to implement in practice, since they may involve searching very large sets of words in the process of error-correction. Much research, some of it highly mathematical, is going on in this relatively new science, which has been in existence only since 1948. If you want to read more on the subject, Linn and Costello (1983) in the 'Suggestions for further reading' would be a good starting point.

PROBLEMS

Problems 1–5 refer to the code with $n = 5$ and $k = 2$, in which the four message words 00, 01, 10, 11 are encoded into 00000, 01101, 10110, 11011. This is a systematic code in which the message bits appear as the rightmost two characters in the valid codeword — that is, as bits x_3 and x_4 using the numbering system we adopted earlier in the chapter. We will explore how this repositioning of the message bits affects the generator and parity check matrices.

1. Can you spot three independent parity relations satisfied by this code. ('Independent' here means that no one of the relations can be derived from the other two. Three relations are needed because there are three parity bits to be determined.)
2. Write the parity relations you have derived in matrix form.
3. Hence write down the generator matrix for the code.
4. What is the minimum distance of this code, and how many errors can it detect and correct?
5. Write down a parity check matrix for the code, and use it to detect and correct the errors in the following received words: (a) 01001; (b) 11010.

Ch. 10] **Problems** 263

6. What is the maximum number of message bits which can be carried by a code with wordlength $n = 11$?
7. Write down a parity check matrix for the three-fold repetition code, and test your answer by checking the received words (a) 000; (b) 001.
8. Examine the following code with $n = 6$ and $k = 3$:

    000000
    001011
    010110
    011101
    100101
    101110
    110011
    111000

    (a) Show that this code can detect and correct single errors.
    (b) Verify that a possible set of parity relations for this code is

    $$x_0 + x_1 = x_3$$
    $$x_1 + x_2 = x_4$$
    $$x_1 + x_3 = x_5 \ .$$

    (c) Hence write down a parity check matrix for the code.
    (d) What happens if you try to use your parity check matrix to check the received word 111111? Why does this occur?

Suggestions for further reading

A book which covers the use of action diagrams in computing is *Action diagrams: clearly structured program design* by James Martin and Carma McClure (Prentice Hall, 1985). The latest edition of James Martin's *Recommended diagramming standards for analysts and programmers* (Prentice Hall, 1987) also contains a section on action diagrams. Both books are relatively easy to read, but rather expensive.

For linear algebra and matrices see *Linear algebra and its applications* by D. H. Griffel (Ellis Horwood, 1989).

A brief treatment of matrices as applied to computer graphics can be found in *Introduction to interactive computer graphics* by Joan E. Scott (Wiley, 1982).

For list processing see *Essential mathematics for software engineers* vol. 4. Ed. Gil Slater (Peter Peregrinus, 1987). This book is part of a videotape/book course, covering various aspects of mathematics, and is supported by the Institute of Electrical Engineers.

Linked lists and other data structures, including graphs and trees, are dealt with in *Discrete mathematical structures in computer science* by Kolman and Busby (Prentice Hall, 1987).

Graphs are covered in *An introduction to graph theory* by R. Wilson (Longman, 1985).

A book which considers trees in the wider context of data structures is *Data structures and algorithms* by Aho, Hopcraft and Ullman (Addison Wesley, 1983). This book is unlikely to become out of date in the foreseeable future!

The book *Database analysis and design* by Hugh Robinson (Chartwell Bratt, 1981) describes a number of database systems and gives an idea of the scope of the subject.

Many books with the word 'probability' in the title tend to be rather mathematical and abstract. *Modern elementary statistics* by Freund (6th edition, Prentice Hall, 1984) is old — but good — and introduces statistics via probability.

Error control coding by Shu Lin and Daniel Costello (Prentice Hall, 1983) covers a wide range of binary coding systems including Hamming codes.

The original paper by Hamming can be found in unit 14 of the Open University third level course TM.361 *Graphs, networks and design* (Open University Press, 1981): and is not difficult to follow.

Answers to exercises and problems

CHAPTER 1
Exercise 1.6.1

```
Entry: an exam mark
    ┌IF mark<0
    │ give warning
    ├ELSE
    │   ┌IF mark<40
    │   │ result='fail'
    │   ├ELSE
    │   │   ┌IF mark<65
    │   │   │ result='pass'
    │   │   ├ELSE
    │   │   │   ┌IF mark<100
    │   │   │   │ result='distinction'
    │   │   │   ├ELSE
    │   │   │   │ give warning
    │   │   │   └ENDIF
    │   │   └ENDIF
    │   └ENDIF
    └ENDIF
Exit: a grade or a warning
```

Problems

1.   Entry:
$$\begin{array}{l} I=0 \\ \text{DO} \\ \quad I=I+1 \\ \quad \text{print } I \\ \text{UNTIL } I=100 \end{array}$$

Exit: list of integers from 1 to 100 inclusive

If the instruction $I=I+1$ is replaced by $I=I+2$ then the output will be the even integers from 2 to 100 inclusive.

2.   Entry: a pre-tax income (i.e., its value)
```
IF pre_tax_income<50
   post_tax_income=pre_tax_income
ELSE
   taxable_income=pre_tax_income-50
   tax_paid=taxable_income×0.2
   post_tax_income=pre_tax_income-tax_paid
ENDIF
```
Exit: post-tax value of income

3. As with all action diagrams, there are many possible answers: the version below checks all possible four-legged animals, then all two-legged animals — which leaves the snake as the only remaining possibility.

```
Entry: an animal
IF number_of_legs=4
   IF colour=pink
      animal is a pig
   ELSE (animal is not pink)
      IF animal has horns
         animal is a cow
      ELSE
         animal is a horse
      ENDIF
   ENDIF
ELSEIF number_of_legs=2
   IF colour=green
      animal is a devil
   ELSE
      animal is a man
   ENDIF
ELSE
   animal is a snake
ENDIF
Exit: animal named
```

Chapter 2

With this diagram the sheep would be identified as a horse. This is because we assumed (correctly for the original set of animals) that if a four-legged animal was not pink, then it had to be brown.

CHAPTER 2

Exercise 2.2.1
(a) Before being able to list all members of this set, you would need to arrive at a definition of what does, and what does not, constitute a valid 'route' from London to Bristol. For instance, is a route via Edinburgh 'valid'?
(b) is definitely acceptable, the set is $\{0, 1, 2, \ldots, 19, 20\}$.
(c) is well-defined — a statement belongs to the set if it does not contain an error (statements could be tested by attempting to compile). The formal specification of a language such as PASCAL will consist of a listing of the members of this set — which is large but finite.
(d) This one would be hard to define; when is an animal 'domestic'?

Exercise 2.3.1
True, false, false, false, true, false.

Exercise 2.3.2
(a) $L = M$
(b) $[S,N]$ $[L,N]$ $[E,N]$ $[M,N]$, $[E,L]$ $[E,M]$, $[L,M]$ $[M,L]$. In addition all sets are subsets of themselves, so that all pairs of the form $[X,X]$ are included.
(c) This is the set of 'fractions', i.e. numbers such as 5/7 or 2/5. With the definition as in the question it also includes fractions such as 4/12 which could be cancelled. The technical term for such numbers is rational numbers.

Exercise 2.5.1
(a) The set of all terminals at the Greenfield site.
(b) The set of all terminals except those at the Fairfield site.
(c) The set of all terminals at either Fairfield or Hilltop — i.e. all except those at Greenfield.
(d) The set of all Engineering terminals.

Exercise 2.5.2
(a) S'
(b) $S' \cap R$
(c) $(D \cup R) - (D \cap R)$
(d) $D \cap S'$

Exercise 2.6.3
The universal set has the required property.

Answers to exercises and problems

Exercise 2.7.1
{a,b,c,d} {a,b,c} {a,b,d} {a,c,d} {b,c,d} {a,b} {a,c} {a,d} {b,c} {b,d} {c,d} {a} {b} {c} {d} ∅

The power set of the three-element set had eight elements; that of the four-element set has 16 elements. It seems as if the power set of an n-element set has 2^n elements.

Exercise 2.9.1
The ordered pairs $[A,B]$ representing 'A can receive a message from B' are: [I,II] [I,III] [I,IV] [I,V] [II,I] [II,III] [II,IV] [II,V] [III,I] [III,II] [III,IV] [III,V].
The ordered pairs for 'can send a message to' can be obtained from those above by reversing the order of the elements.

Exercise 2.9.2
$M \times N = \{[1,0], [1,1], [2,0], [2,1], [3,0], [3,1], [4,0], [4,1], [5,0], [5,1]\}$.

Exercise 2.10.1
⟨M,P,U,T,E⟩,O

Exercise 2.10.2
⟨D,I,S,P,U,T,E⟩

Exercise 2.10.3
Start = 1

Address	Day name	Pointer
1	Sunday	2
2	Monday	6
3	Wednesday	7
4	Friday	5
5	Saturday	0
6	Tuesday	3
7	Thursday	4

Problems
1. Define the set of students eligible to take A11=EA11, and similarly for those eligible to take the other two subjects. In the same way, define the set of students who passed I5 as PI5, etc, and the set of students obtaining credit in I5 as CI5, etc. Note that CI5⊂PI5 and so on.

Chapter 3

Then if X denotes a student, we can say
    if $X \in (CI5 \cap PI8) \cup (PI5 \cap CI8C)$ then $X \in EA11$
    if $X \in PI5 \cap PI8$ then $X \in EI11$
    if $X \in PI5 \cap (CI4 \cup CI2)$ then $X \in EA12$

2. Read list L1
    List L2 = $\langle \; \rangle$
    ⎡While L1 $\langle \; \rangle$
    ⎢ C1C2C3 = Head (L1)
    ⎢ If C1 = P then L2 = append(L2, $\langle$C1C2C3$\rangle$)
    ⎢ L1 = tail (L1)
    ⎣Endwhile
    End
(This is only one of several possible solutions)

3. One possible representation would be to use ordered triples, thus: $[X,Y,Z]$, where $X \in \{\text{depots}\}$, $Y \in \{\text{plants}\}$, and $Z = $ quantity. Then plants supplied only by depot C could be identified by using an algorithm of the form
    Read X
    If $X = C$ then add Y to list
    etc.

CHAPTER 3

Exercise 3.3.1
$P(0) = F$, $P(1) = F$, $P(2) = P(4) = P(8) = T$, $P(16) = F$.

Exercise 3.4.1

a	b	$a \lor b$
T	T	T
T	F	T
F	T	T
F	F	F

Exercise 3.4.3
$P(12) \lor Q(9) = T$, $P(9) \lor Q(12) = T$.

Exercise 3.4.4
If $x = $ Queen Elizabeth I, $P(x) \lor Q(x) = T$, $P(x) \land Q(x) = T$
If $x = $ Charles Babbage, $P(x) \lor Q(x) = T$, $P(x) \land Q(x) = F$
If $x = $ Joan of Arc, $P(x) \lor Q(x) = T$, $P(x) \land Q(x) = F$
If $x = $ Albert Einstein, $P(x) \lor Q(x) = F$, $P(x) \land Q(x) = F$.

Exercise 3.5.1

A	B	A∧B	$\overline{A \wedge B}$	$\overline{A}$	$\overline{B}$	$\overline{A} \vee \overline{B}$
T	T	T	F	F	F	F
T	F	F	T	F	T	T
F	T	F	T	T	F	T
F	F	F	T	T	T	T

Exercise 3.5.3
$(\overline{A \wedge B})$ means 'x is not an even number divisible by three'. $\overline{A} \vee \overline{B}$ means 'x is not even, or it is not divisible by three, or both'. These two statements are equivalent.

Exercise 3.7.1
(a) $\overline{F(x)} \wedge \overline{A(x)} \wedge B(x)$
(b) $D(x) \wedge \{[A(x) \wedge \overline{F(x)}] \vee [\overline{A(x)} \wedge F(x)]\}$

Exercise 3.9.1
$S(x) \Rightarrow P(x)$. To qualify, it is necessary to pass the exam; a sufficient condition to have passed the exam is to have qualified.
 $S(x) \Rightarrow Q(x) \vee R(x)$. To qualify, it is necessary either to get over 40% in all assessments, or to get over 45% in all except one. To have achieved either over 40% in all assessments, or over 45% in all except one, it is sufficient to have qualified.
 $S(x) \Leftrightarrow P(x) \wedge (Q(x) \vee R(x))$

Exercise 3.9.2

⇒	H	N	E	O
H	T	F	F	T
N	T	T	T	T
E	T	T	T	T
O	F	F	F	T

Thus E and N form a pair of necessary and sufficient conditions.

Exercise 3.9.3
(a) T
(b) T
(c) F

Problems
1. (a) Constructive proof.
   (b) Induction.
   (c) Contradiction.

3. (a) 1, 3, 6.
   (b) $0.5n(n-1)$.

4. (a) (i) $B(x) \wedge A(x) = T$.
       (ii) $D(x) \wedge \overline{B}(x) \wedge A(x) = T$.
       (iii) $\overline{A}(x) \vee \overline{D}(x) = T$.
   (b) F, T, F.

5. (a)

A	B	$A \vee B$	$\overline{A \vee B}$	C	$\overline{(A \vee B)} \Rightarrow C$
T	T	T	F	T	T
T	F	T	F	T	T
F	T	T	F	T	T
F	F	F	T	T	T
T	T	T	F	F	T
T	F	T	F	F	T
F	T	T	F	F	T
F	F	F	T	F	F

(b) Yes.

CHAPTER 4

Exercise 4.5.1
19985

Exercise 4.5.2
110111

Exercise 4.7.1
9

Exercise 4.7.2
Numbers of the form $6k + 4$ are those which are congruent to 4 (mod 6).

Exercise 4.8.2
(a) 6.3.
(b) 6.4.
(c) 6.2.

Exercise 4.10.1
(i) is square
(ii) is square, symmetric
(iii) has none of these characteristics
(iv) is square and diagonal.

Exercise 4.10.2
(i) $a_{22} = 2$.

(ii) $A' = \begin{bmatrix} 2 & 3 & 7 \\ 4 & 2 & 10 \end{bmatrix}$.

Exercise 4.12.1
$\begin{bmatrix} 0 & 2 \\ 4 & 1 \\ 6 & 3 \end{bmatrix}$

Exercise 4.12.2
$\begin{bmatrix} 2a & 2b \\ 2c & 2d \\ 2e & 2f \end{bmatrix}$

The multiplication has the effect of multiplying each element in the second matrix by two.

Exercise 4.12.3
$\begin{bmatrix} 0 \\ 0 \end{bmatrix}$

The multiplication gives a zero matrix as the result irrespective of the values of x and y.

Exercise 4.12.4
$\begin{bmatrix} 6 & 7 \\ 8 & 6 \end{bmatrix}, \begin{bmatrix} 2 & 8 \\ 5 & 10 \end{bmatrix}$

The result depends on the order in which the multiplication is carried out.

Exercise 4.13.1
Multiply by
$\begin{bmatrix} -1 & 0 \\ 0 & 1 \end{bmatrix}$

Exercise 4.13.2
Magnification by 2 plus reflection in the x-axis.

Exercise 4.13.3
Multiply by

$$\begin{bmatrix} 0 & 1 \\ -1 & 0 \end{bmatrix}$$

Problems
1. ENTER
    READ 24-HOUR CLOCK TIME T
    $\quad$ IF $T < 12$
    $\quad\quad$ PRINT "TIME IS" T "A.M."
    $\quad$ ELSE
    $\quad\quad$ IF $T = 12$
    $\quad\quad\quad$ PRINT "TIME IS NOON"
    $\quad\quad$ ELSE
    $\quad\quad\quad$ PRINT "TIME IS" $T - 12$ "P.M."
    $\quad\quad$ ENDIF
    $\quad$ ENDIF
    EXIT

2. About 8.5%; about 1% (this is assuming that figures are truncated rather than rounded).

3. (a) $\begin{bmatrix} 6 & 4 & 4 \\ 5 & 6 & 1 \end{bmatrix}$

   (b) $G = \begin{bmatrix} 1.50 \\ 1.50 \\ 2.00 \end{bmatrix}$ $U = \begin{bmatrix} 1.50 \\ 0.10 \\ 1.20 \end{bmatrix}$

   (c) $G_1 = \begin{bmatrix} 0.51 \\ 0.51 \\ 0.68 \end{bmatrix}$ $U_1 = \begin{bmatrix} 0.30 \\ 0.06 \\ 0.73 \end{bmatrix}$

   (d) $EG_1 = \begin{bmatrix} 7.82 \\ 6.29 \end{bmatrix}$ $EU_1 = \begin{bmatrix} 4.96 \\ 2.59 \end{bmatrix}$

   These two matrices contain the total prices for assembling the two prducts using German and US components respectively. Clearly the US alternative is cheaper.

4. (a) The square becomes a rhombus (parallelogram with equal sides) with side length 7 and corners at [0,0], [7,0], [13,7], [6,7].
   (b) The z co-ordinates of the points remain unchanged, while the x and y co-ordinates are interchanged and multiplied by two. The cube thus becomes a cuboid, height 1, length and breadth 2.

5. 25 hours mixing, 100 hours heating and 80 hours setting.

6. (a) 24
   (b) $m \times n \times p$
   (c) Better to multiply the 8×5 and the 5×2 first, then multiply the result by the 3×8 — this only gives 128 multiplications, as against 150 if we begin by multiplying the first two matrices.

7. (a) ENTER $m \times n$ matrix A
   $k = 0$
   $$\begin{array}{l} \vdash \text{DO WHILE } i < m \\ \quad \vdash \text{DO WHILE } j < n \\ \quad \quad k = k + 1 \\ \quad \quad R_k = A_{ij} \\ \quad \vdash \text{endwhile} \\ \vdash \text{endwhile} \end{array}$$
   EXIT

   (b) ENTER ROW MATRIX WITH $m \times n$ ELEMENTS
   $k = 0$
   $$\begin{array}{l} \vdash \text{DO WHILE } i < m \\ \quad \vdash \text{DO WHILE } j < n \\ \quad \quad k = k + 1 \\ \quad \quad A_{ij} = R_k \\ \quad \vdash \text{endwhile} \\ \vdash \text{endwhile} \end{array}$$
   EXIT

CHAPTER 5

Exercise 5.2.1
Both are valid graphs, though the second one is disconnected (see Section 5.4).

Exercise 5.3.1
The first two graphs are isomorphic. One correspondence of vertices is $[c,v]$, $[d,w]$, $[e,x]$, $[a,y]$, $[b,z]$.

In the case of the second pair, although the two graphs have identical degree sequences $\langle 3,2,2,2,1 \rangle$, they are not isomorphic.

Exercise 5.8.1
(a) The vertices a, b, g, and e are the vertices of a complete subgraph.
(b) To get a set of complete subgraphs which between them contain all the edges of G we can include the additional three triangles with vertex sets $\{b,c,e\}$, $\{c,d,e\}$ and $\{e,f,g\}$ respectively.

Chapter 5

Exercise 5.8.2
For a K_5 subgraph to be present, there would need to be at least five vertices of degree at least 4: similarly for a $K_{3,3}$ we would need six vertices of degree at least 3.

Exercise 5.8.3
One possibility is the K_4 graph with vertices a, b, c and d, together with the connected graph consisting of edges $\{a,e\}$, $\{b,e\}$, $\{c,e\}$ and $\{d,e\}$.

Problems
1. The graphs G_1, G_2 and G_3 are shown in Figs. S.1, S.2 and S.3.

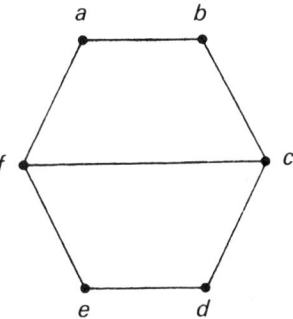

Fig. S.1.

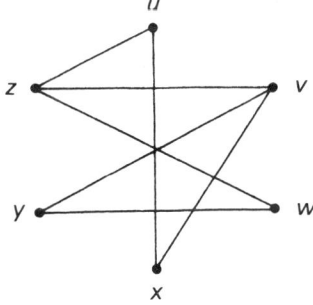

Fig. S.2.

2. The degree sequences are $\langle 3,3,2,2,2,2 \rangle$, $\langle 3,3,2,2,2,2 \rangle$, and $\langle 3,2,2,2,1,1,1 \rangle$.
4. Note that the bipartition of the original vertex set gives the two subsets $\{u,v,w\}$ and $\{x,y,z\}$.

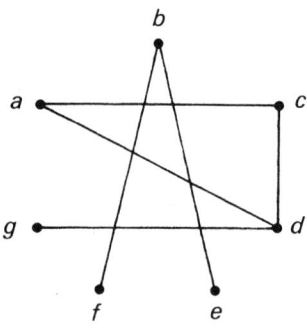

Fig. S.3.

5. A longest circuit in G_1 is $\langle a,b,c,d,e,f,a \rangle$.
   Two shortest circuits are $\langle a,b,c,f,a \rangle$ and $\langle f,c,d,e,f \rangle$.
6. The degree sequences of G_1 and G_2 are the same, so the two graphs *could* be isomorphic. We can make a correspondence between the vertices which have three incident edges in G_1 (vertices c and f) and the vertices which have three incident edges in G_2 (vertices v and z). Further, since (in G_1) the three adjacent vertices to c are b, d and f, we can check the equivalent adjacent vertices to v (in G_2) — these adjacent vertices are x, y and z (we can check the adjacent vertices of the other 'three-edge' vertices, f and z, in similar fashion). The tentative correspondence between the vertices of G_1 and the vertices of G_2 is shown in Fig. S.4. Now we must check whether edge-pairs correspond as well. Verify from

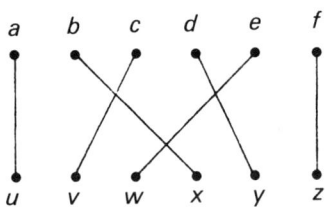

Fig. S.4.

Fig. S.4 that the edges of G_1 have a one-to-one correspondence with the edges of G_2 (for a start, edge $\{a,b\}$ corresponds to edge $\{u,x\}$ just as vertex a corresponds to vertex u, and vertex b corresponds to vertex x).

7. (a) Two rooms will be needed. Suppose we colour vertex f *red*: vertices a, e and c could be coloured *blue*, leaving b and d to be coloured red. So a satisfactory allocation of staff is Andy, Clare and Eddy in one room, and Barry, Dave and Frank in the other.
   (b) If we include Gerry — by the inclusion of vertex g together with incompati-

bility edges $\{a,g\}$ and $\{b,g\}$ — then we get a triangle in the graph, made up of vertices a, b and g. This means that we will need three vertex colours and consequently three rooms will be necessary.

(c) The *compatibility* graph G_4 will consist of the vertex set V_4, where $V_4 = \{a,b,c,d,e,f,g\}$ and the set of edge pairs E_4 where
$E_4 = \{\{a,c\},\{a,d\},\{a,e\},\{b,d\},\{b,e\},\{b,f\},\{c,e\},\{c,g\},\{d,f\},\{d,g\},$
$\{e,g\},\{f,g\}\}$.

(d) The graph is shown in Fig. S.5.

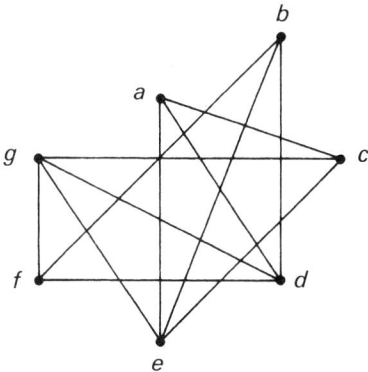

Fig. S.5.

One possible minimum collection of complete subgraphs which includes all vertices is the collection of three graphs on the vertices $\{a,c,e\}$, $\{b,d,f\}$ and $\{d,f,g\}$ respectively. An allocation of people to rooms is a, c and e in one room: b and d in another; and f and g in the third. Notice that the amiable Frank and Dave have alternative choices (though to put them in the same room would require a total of four rooms).

CHAPTER 6

Exercise 6.1.1
The digraph is
$G = [V,E]$
$V = \{be, fx, gn, gs, mo, ra\}$
$E = \{[be,fx], [be,mo], [be,ol], [be,ra], [fo,mo],$
$[fo,ra], [mo,gn], [ol,be], [ol,mo], [ra,gs]\}$.

Problems

1. (a) The digraph is drawn in Fig. S.6.
   (b) The ordered pair representation is
   $G = [V,E]$
   $V = \{a, b, c, d\}$
   $E = \{[a,b], [a,d], [b,b], [b,c], [b,d], [d,c]\}$

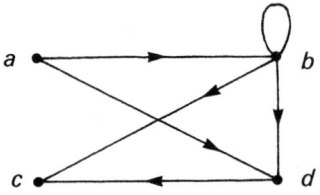

Fig. S.6.

(c) A simple linked-list representation is
start 1

address	record	pointer
1	[a,b]	2
2	[a,d]	3
3	[b,b]	4
4	[b,c]	5
5	[b,d]	6
6	[d,c]	0

The list could be modified as follows to produce an efficient list
start $\langle 1,4,2,0 \rangle$

address	record	pointer
1	[a,b]	3
2	[a,d]	5
3	[b,b]	0
4	[b,c]	6
5	[b,d]	0
6	[d,c]	0

(d) To amend the matrix version we change the zero entry in row 3, column 1 to a value of 1. In the case of the pictorial version of the graph we can simply draw a directed edge from c to a. To update the ordered pair representation we can include the extra ordered pair $[c,a]$ as an element of the set E. The first linked list would have a modified pointer from address 6, and a new record in address 7, as shown below.

$$\begin{array}{lll} 6 & [d,c] & 7 \\ 7 & [c,a] & 0 \end{array}$$

In the 'efficient' linked-list version, address 7 should be accessed first as the head vertex is first in alphabetic order; so the new version of this linked list has a modified 'start list' — and the same additional record in address 7 as above. The new boss list is $\langle 7,1,4,2,0 \rangle$

(e) Similar arguments apply if a new vertex is included, though in programming terms it might be awkward to change the dimensions of a matrix by the addition of a complete new row and column. Note that the 'efficient' linked list would have a new boss list $\langle 7,1,4,2,8,0 \rangle$ and a new record in address 8 (assuming that we leave the recently included record at address 7). The record will be 8 $[c,e]$ 0.

2. (a) The digraph is shown in Fig. S.7.

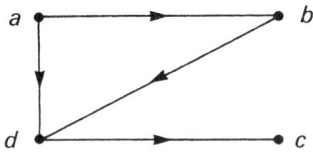

Fig. S.7.

(b) If we call the matrix M, the matrix logical product $M \times M$ is obtained as follows

$$\begin{array}{c} \begin{array}{cccc} a & b & c & d \end{array} \\ \begin{array}{c} a \\ b \\ c \\ d \end{array} \left[\begin{array}{cccc} 0 & 1 & 0 & 1 \\ 0 & 0 & 0 & 1 \\ 0 & 0 & 0 & 0 \\ 0 & 0 & 1 & 0 \end{array} \right] \\ M \end{array} \times \begin{array}{c} \begin{array}{cccc} a & b & c & d \end{array} \\ \left[\begin{array}{cccc} 0 & 1 & 0 & 1 \\ 0 & 0 & 0 & 1 \\ 0 & 0 & 0 & 0 \\ 0 & 0 & 1 & 0 \end{array} \right] \\ M \end{array} = \begin{array}{c} \begin{array}{cccc} a & b & c & d \end{array} \\ \left[\begin{array}{cccc} 0 & 0 & 1 & 1 \\ 0 & 0 & 1 & 0 \\ 0 & 0 & 0 & 0 \\ 0 & 0 & 0 & 0 \end{array} \right] \begin{array}{c} a \\ b \\ c \\ d \end{array} \\ M^2 \end{array}$$

so there are paths of length 2 from a to c; from a to d; and from b to c.

(c) Take as an example $M^2[1,3]$ with value 1. The result is obtained from

$$
\begin{array}{c}
\ a\ b\ c\ d \\
a\ [0\ 1\ 0\ 1] \\
*\times
\end{array}
\begin{array}{c}
c \\
\begin{bmatrix} 0 & a \\ 0 & b \\ 0 & c \\ *1 & d \end{bmatrix}
\end{array} = 1
$$

The two marked elements indicate the existence of an edge from a to d, and an edge from d to c. This means that there is a path of length 2 from a to c. Also, the logical product is 1 (or *true*) because the two marked elements both have value 1.

(d) The only entry of value 1 in the matrix M^3 is in row 1 column 3, indicating a path of length 3 from a to c.

(e) $M^n = 0$ for $n > 3$.

(f) A loop on vertex b would give a perisistent 1 in the second diagonal element, no matter how many multiplications were carried out. There would be paths of all lengths from b to b; from a to d; and paths of length >1 from a to c.

3. (a) The diagraph is shown in Fig. S.8.

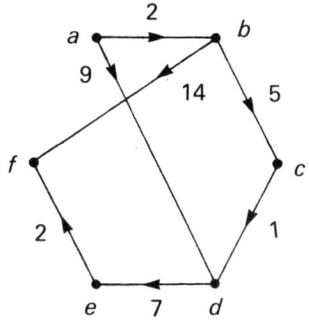

Fig. S.8.

(b) Five iterations are required. The value M represents an initial very large vertex label. An asterisk is attached to a label when the label becomes permanent: the appropriate backward reference is entered to the right of the label — whether or not the label is temporary or permanent.

Iteration	Vertex					
	a	b	c	d	e	f
0 (i)	0	M	M	M	M	M
0 (ii)	*0	M	M	M	M	M
1 (i)	*0	2(a)	M	9(a)	M	M
1 (ii)	*0	*2	M	9	M	M
2 (i)	*0	*2	7(b)	9	M	16(b)
2 (ii)	*0	*2	*7	9	M	16
3 (i)	*0	*2	*7	8(c)	M	16
3 (ii)	*0	*2	*7	*8	M	16
4 (i)	*0	*2	*7	*8	15(d)	16
4 (ii)	*0	*2	*7	*8	*15	16
5 (i)	*0	*2	*7	*8	*15	16
5 (ii)	*0	*2	*7	*8	*15	*16(b)

The reverse path is $\langle f,b,a \rangle$, so the minimum weight is $\langle a,b,f \rangle$ (with weight 16).

Notice that there is a change to the value of the temporary label on vertex d during iteration 3: notice also that in iteration 5(i) the possible labelling of vertex f as 17(e) is rejected as f already has a (temporary) label of value 16.

4. (a) The precedence graph is shown in Fig. S.9 with earliest start and finish times marked.

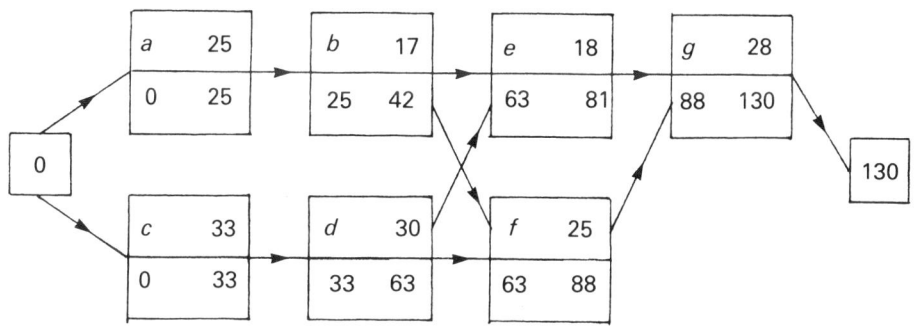

Fig. S.9.

(b) The audit will finish at 16 minutes after midnight. No file/ledger/book will be in use throughout the audit (though the cash file is required during all but 21 minutes).

CHAPTER 7

Exercise 7.7.1
The tree has a maximum of four levels: you should find that the path from the root (96) to the vertex (150) is R, R, L, stop.

Exercise 7.7.2
See the answer to Question 9 for a brief discussion.

Exercise 7.7.3
We could amend the rule 'if the value of the new element is less than {or equal to} the vertex value go left'. We might also wish to amend the algorithm further to record the existence and value of ties.

Problems
1. (a) See Fig. S.10.

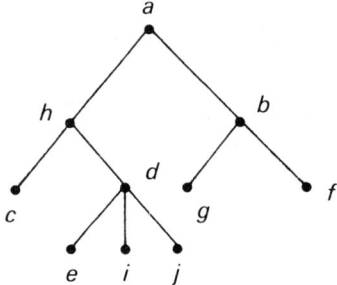

Fig. S.10.

(b)
$$\begin{bmatrix} a \\ \begin{bmatrix} h \\ c \\ \begin{bmatrix} d \\ \begin{bmatrix} e \\ i \\ j \end{bmatrix} \end{bmatrix} \end{bmatrix} \\ \begin{bmatrix} b \\ \begin{bmatrix} g \\ f \end{bmatrix} \end{bmatrix} \end{bmatrix}$$

(c) $\langle a \langle h \langle c \rangle \langle d \langle e \rangle \langle i \rangle \langle j \rangle \rangle \rangle \langle b \langle g \rangle \langle f \rangle \rangle \rangle$

2. We have a problem! The zero at the end of 510 is read as a number. It would be more logical to treat the 510 as 51-; then the vertices 510, 511 and 516 would all be children of 51-. The vertex 510.9 could then be placed at the right level.

3. (a) Se Fig. S.11.

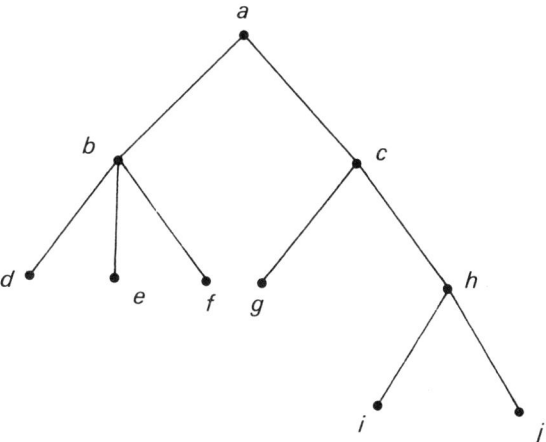

Fig. S.11.

(b) $\begin{bmatrix} a \\ \begin{bmatrix} b \\ d \\ c \\ f \end{bmatrix} \\ \begin{bmatrix} c \\ g \\ \begin{bmatrix} h \\ i \\ j \end{bmatrix} \end{bmatrix} \end{bmatrix}$

4. (a) See Fig. S.12.

5. The edges selected in order are, by initial of vertex and with weights: $\{b,e\}$, 3; $\{b,f\}$, 4; $\{a,d\}$, 7; (reject $\{e,f\}$, 7); $\{a,c\}$, 9; $\{b,d\}$, 9. Frankfurt and Carthage are furthest apart — a distance of 29 hundred miles, compared with a direct distance of 13 hundred miles.

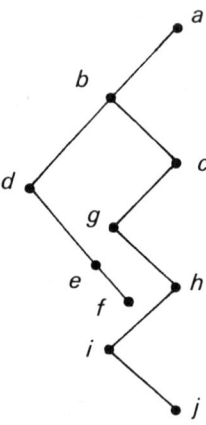

Fig. S.12.

6. Suppose we start from vertex a. The choice of edges in order, with weights, is $(a,c), 5; (c,b,), 5; (b,d), 4; (d,e), 4; (b,f), 8$. Note that these pairs are ordered: we look from the first element to the second. If all machines communicate approximately the same amount of information, then machine b will be the busiest, as it is involved with all communications except those between the pairs $\{e,d\}$ and $\{a,c\}$.

7. (a) See Fig. S.13.

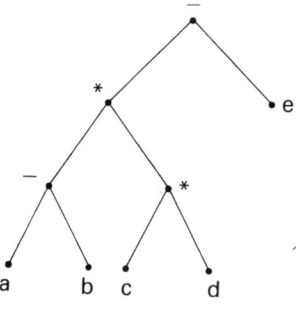

Fig. S.13.

(b) The pre-order search gives the expression
$- * - a\ b * c\ d\ e$ and the post-order gives
$a\ b - c\ d * * e -$
The first evaluation is
$- * - 3\ 2 * 5\ 10\ 2$
$- * \quad\quad 1 \quad\quad 50\ 2$
$- \quad\quad 50 \quad 2$
$\quad\quad 48$

The second is
3 2 − 5 10 * * 2 −
   1   50   * 2 −
          50 2 −
              48

8. (a) Four of the states in the process are shown (note that all numbers are single digit numbers).

Output	Stack	Input
		⊥3+4*(7−5) ↑2⊥
3475−	(	) ↑2⊥
	*	
	⊥	
3475−2	↑	⊥
	*	
	+	
	⊥	
3475−2 ↑*+	⊥	⊥

The evaluation in sequence is
3 4 7 5 − 2 ↑ * +
3 4 2 − 2 ↑ * +
3 4   4   * +
3   16   +
19

(b) The tree is shown in Fig. S.14.

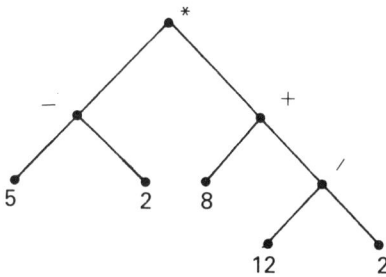

Fig. S.14.

(c) The preorder search gives
*−5 2+8 / 12 2   which can be evaluated as shown
*   3 +8   6
*   3   14
     42

9. (a) The tree is shown in Fig. S.15.

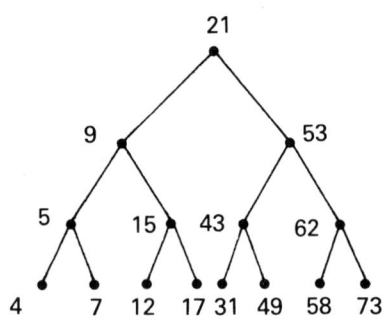

Fig. S.15.

(b) The sorted list is
4, 5, 7, 9, 12, 15, 17, 21, 31, 43, 49, 53, 58, 62, 73

(c) You could investigate the fact that 21 (the first numnber in the original list) is the middle value of the ordered list. On the other hand, an already sorted list would give rise to a tree of 15 levels.

CHAPTER 8

Exercise 8.4.1

(a)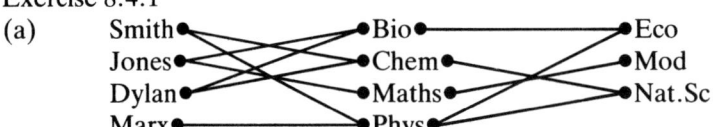

(b) The composition $R_1 \circ R_2$ is

Smith — Eco
Jones — Mod
Dylan — Nat.Sc
Marx

	Eco	Mod	Nat.Sc
Smith	1	0	1
Jones	1	1	0
Dylan	1	0	1
Marx	1	0	1

$R_1 \circ R_2$ = {[Sm,Na], [Sm,Ec], [Jo,Ec], [Jo,Mo], [Dy,Ec], [Dy,Na], [Ma,Ec], [Ma,Na]}

The set of triples is

{[Sm,Ch,Na], [Sm,Ph,Ec], [Sm,Ph,Na], [Jo,Bi,Ec], [Jo,Ma,Mo], [Dy,Bi,Ec], [Dy,Ch,Na], [Ma,Ph,Ec], [Ma,Ph,Na]}

Note that there are two triples corresponding to the ordered pair [Sm,Na].

(c) The existence of an ordered pair $[x,y]:[x,y] \in R_1 \circ R_2$ means that student x has *some* of the pre-requisites for second level subject y.

Problems

1. (a) The ordered pair representation is {[3,7],[5,7],[7.7],[10,4],[10,6]}
   (b)
   3● ●4
   5● ●6
   7● ●7
   10●
   (c) The relation S is reflextive as $|a-a|=0$, and 0 divides by n exactly 0 times.
   The relation is symmetric: If $|a-b|$ divides by n, then so does $|b-a|$. Suppose $|a-b|$ divides exactly by n and that $|b-c|$ also divides exactly by n. Then $(a-b)$ and $(b-c)$ will also divide exactly by n (the result could be a negative integer, but that doesn't matter). This means that $(a-b)+(b-c)$ will also divide exactly by n, since each bracketed term must be a multiple of n.
   Since $(a-c)$ will divide exactly by n, so will $|a-c|$; we see that the relation is transitive.

2. (a)

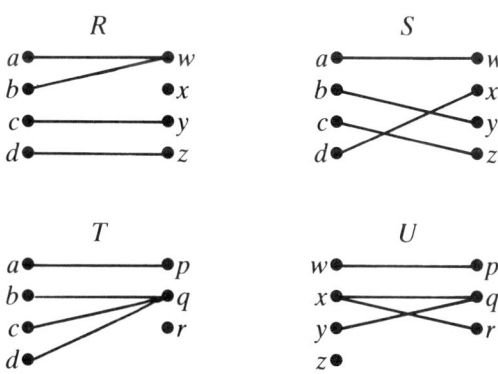

(b) R, S and T are functions as each object element has a single corresponding

image. U is not a function: firstly there is no image corresponding to the object z; secondly the object x has two images, q and r.
(c) S is the only reversible function (that is, a function with an inverse).

3. (a) To keep the picture clear, only five of the 99 elements of Q are shown in the diagram.

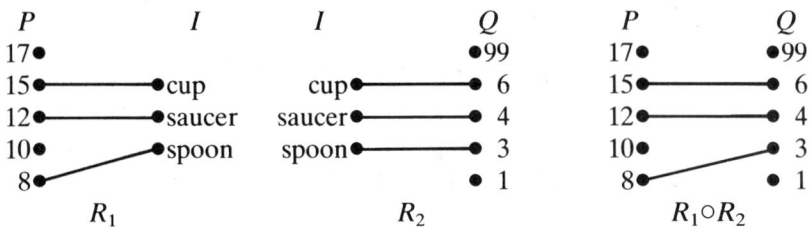

(b) The ordered triples are [15,cup,6], [12,saucer,4] and [8,spoon,3].
(c) If we consider elements of P, I and Q as unit_price, item and quantity respectively: and if we consider the relations R_1 and R_2 to be the files Price_file and Order_file, then a plausible query would be

<u>FROM</u> Price_file, Order_file
<u>SELECT</u> unit_price, quantity
<u>WHERE</u> item <u>IN</u> Price_file = item <u>IN</u> Order_file

4. (a) $M = P(A) = \{\emptyset, \{f,c,p\}, \{f,c\}, \{f,p\}, \{c,p\}, \{f\}, \{c\}, \{p\}\}$.

(b) $B = \{1, 1.5, 2, 2.5, 3, 3.5, 4\}$.

(c) R_1 (relation of person to meal)

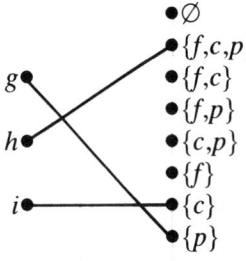

(d) R_2 (relation of meal to price)

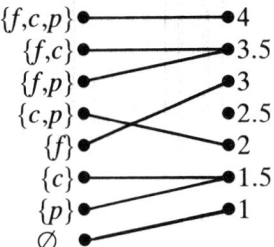

Chapter 8

(e) The tabulation of the triples is

g   p   1.5
h   f,c,p   4
i   c   1.5

(f) Both relations are functions, and so is the composition. Notice that the apparent one-to-one nature of R_1 is accidental: none of the diners ordered the same meal on this particular occasion!

5. (a) $M \times X$ consists of 66 ordered pairs.
   16 of them are in relation A.
   (b) The ordered pairs of set D are tabulated
   M4 London
   M3 London
   M2 London
   M1 London
   M4 Bath
   M4 Bristol
   M4 Cardiff.

   (c) The bipartite graph of relation B is

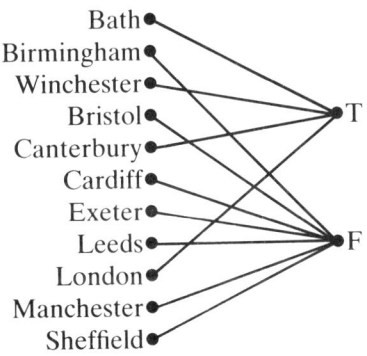

   (d) The cartesian product $C \times H$ contains 22 ordered pairs.
   (e) The composition K consists of 11 ordered pairs, one ordered pair corresponding to each city.
   (f) $V = \{[M4,London,T],[M4,Bath,T]\}$
   (g) $F = \{y : [x,y,z] \in K \text{ and } z = T\}$

6. (a) This raises the case of a legal definition: in the U.K. each private vehicle has a single keeper. If we allow that the term 'keeper' means the same as 'owner' then we can keep the question simple.

   R_1 is not a function, since a given owner could own more than one vehicle;

that is, we can map the name of an owner onto more than one registration number (each registration number is unique).
R_2 is a function.
R_3 is a function: each car has a unique make (a car which is not of a standard make is classified under the catch-all make of 'special').

(b) The composition will consist of a set of ordered pairs such that any ordered pair $[a,b]$ will correspond to the information that person a owns at least one car of make b.

7. (a) Some preconditions are — the inputs should be integer of not more than five digits: the second figure should be greater than the first. Notice that in some circumstances the second figure *could* be both valid and less than the first, for example a mile counter will eventually go 'round the clock' and return to zero, and it might be a requirement to allow for the possibility if, for example, the program related to a safety-critical system. The post-condition is that the output should be a positive integer of five or fewer digits.
(b) The input precondition is that the number should be a positive integer number with exactly n digits. The post-condition is that the output should be a real number (what about the case of a square root which is an integer)? Another possible post-condition is that the output value should be smaller than the input value.
(c) A precondition is that the lengths should be given as real numbers, and that the sum of two smallest numbers should be greater than the largest number. A post-condition on the output is that all angles should be positive, and should sum to two right angles.

CHAPTER 9

Exercise 9.3.1
$8 \times 7 = 56$

Exercise 9.3.2
$5 \times 4 + 3 \times 2 = 26$.

Exercise 9.3.3
(a) $3 + 1 = 4$.
(b) $5 \times 4 = 20$.

Exercise 9.3.4
(a) $24 \times 23 \times 10 \times 9 \times 8$.
(b) $24^2 \times 10^3$.

Exercise 9.3.5
10.

Chapter 9

Exercise 9.4.1
$n=10$, $r=3$, no repeats, order doesn't matter.

Exercise 9.4.2
$n=200$, $r=2$, no repeats, order doesn't matter.

Exercise 9.4.3
$n=5$, $r=3$, no repeats, order matters.

Exercise 9.5.1
$2^4 = 16$

Exercise 9.5.2
16. In the truth table, each of the four components A to D may be either T or F; in the selection of subsets, each of the four elements may be either selected or not selected. If we identify 'select' with T and 'do not select' with F, then the two problems have identical structure.

Exercise 9.5.3
$4 \times 3 \times 2 = 24$.

Exercise 9.5.4
$9!$

Exercise 9.5.5
$9!/2^3$ (because each pair of identical letters which occurs (two r's, two s's and two o's) can be arranged in two different ways).

Exercise 9.5.6
$12!/10!2! = 66$.

Problems
1. 35
2. 0.993, 0.999.
3. 36 (select two person-days from nine, no repetitions, order doesn't matter).
4. (a) $_{11}C_6$ — the path from home to college has 11 steps altogether, of which 6 must be northwards (and the other 5 westwards). So the number of routes is the number of ways of selecting the six northward steps out of the full 11. (This argument of course supposes that only 'sensible' routes — those which do not involve going outside the 6×5 area, or doubling back on one's tracks — are included in the count.)
   (b) $_5C_3 \times _6C_3$ — the path is split into two stages, from home to the corner, and from the corner to college.
5. $10^6 - 1$ (a completely blank key is not allowed).

6. 0.9871299 (this is the probability that one or both lines from A to B are working and one or both lines from B to C are working and CD is working).
7. 3, because the probability of all three breaking down at once is 0.008 which is less than 1%.
8. 12^{256}.
9. (a) Eight.
   (b) 0.088, to three decimal places.
10. (a) 36.
    (b) 30.

CHAPTER 10

Exercise 10.3.1
The check-digit is 4, so the full ISBN is 0-8018-1916-4.

Exercise 10.3.2
(a) is a valid number,
(b) is not, because the weighted sum of the digits is 160, which is not congruent to zero mod 11.

Exercise 10.3.3
One possible example: if the number 0-8914-7328-9 is mis-copied as 0-8914-7832-9, the error will not be detected by the check-digit procedure because the combined weights of the mis-copied digits $8\times4+3\times3+2\times2=45$, while for the correct version $3\times4+2\times3+8\times2=34$, and $34 \equiv 45 \pmod{11}$.

Exercise 10.4.1
$2^5 = 32$.

Exercise 10.4.2
To encode $26+10+24=60$ symbols, as a minimum we would need words of length 6, since $2^6 = 64$.

Exercise 10.5.1
0010.

Exercise 10.5.2
10100.

Exercise 10.6.1
For a four-fold repetition code up to two errors can be detected, but only one corrected (if 0000 is transmitted, and received as 0011, we can detect that there are two errors, but cannot tell whether the transmitted word was 0000 or 1111).

Chapter 10

Exercise 10.6.2
For a five-fold repetition code two errors can be both detected and corrected.

Exercise 10.7.1
Minimum distance $= 2$. For the answer to the second part of the question, see Section 10.9.

Exercise 10.7.3
$$G = \begin{bmatrix} 1 & 0 & 0 & 1 & 1 & 0 \\ 0 & 1 & 0 & 1 & 0 & 1 \\ 0 & 0 & 1 & 0 & 1 & 1 \end{bmatrix}.$$

The encoded words are 010101, 110011.

Exercise 10.8.1
$[1 \ 0 \ 1]$.

Exercise 10.8.2
(a) $S = [0 \ 0 \ 0]$, so word is valid.
(b) $S = [0 \ 0 \ 1]$, so last bit is in error — the transmitted word was 0000000.
(c) $S = [0 \ 1 \ 1]$, so first bit is in error — the transmitted word was 1001100.

Problems

1.
$$\begin{aligned} x_0 &= x_3 \\ x_1 &= x_4 \\ x_2 &= x_3 + x_4. \end{aligned}$$

2.
$$\begin{bmatrix} x_0 \\ x_1 \\ x_2 \end{bmatrix} = \begin{bmatrix} 1 & 0 \\ 0 & 1 \\ 1 & 1 \end{bmatrix} \begin{bmatrix} x_3 \\ x_4 \end{bmatrix}.$$

3.
$$\begin{bmatrix} 1 & 0 & 1 & 1 & 0 \\ 0 & 1 & 1 & 0 & 1 \end{bmatrix}.$$

4. Minimum distance $= 3$; one error can be detected and corrected.

5.
$$H = \begin{bmatrix} 1 & 0 & 0 & 1 & 0 \\ 0 & 1 & 0 & 0 & 1 \\ 0 & 0 & 1 & 1 & 1 \end{bmatrix}.$$

(a) $S = [0 \ 0 \ 1]$. Matches third column of H, so third bit is in error — transmitted word was 01101.
(b) $S = [0 \ 1 \ 1]$. Matches last column of H, so last bit is in error — transmitted word was 11011.

6. Using the relation $2^n/(1+n) > 2^k$ gives maximum $k = 7$.

7.
$$H = \begin{bmatrix} 1 & 0 & 1 \\ 0 & 1 & 1 \end{bmatrix}.$$
   (a) Word is correct.
   (b) $S = [1 \ 1]$. Third bit is in error — word should be 000.

8. (a) As the weight of the lowest-weight codewords is 3, the minium distance is 3 also, so one error can be detected and corrected.
   (c)
$$H = \begin{bmatrix} 1 & 1 & 0 & 1 & 0 & 0 \\ 0 & 1 & 1 & 0 & 1 & 0 \\ 1 & 0 & 1 & 0 & 0 & 1 \end{bmatrix}.$$
   (d) $S = [0 \ 0 \ 0]$. This does not match any column of H, so the algorithm fails — there is more than a single error in the received word.

Index

action diagram, 16
addition principle, 222
address, 44
algorithm, 14
   greedy, 169
   Kruskal's, 172, 177
   Prim's, 173
'and' rule, 222, 227
append function, 43
arithmetic
   floating point, 78
   modulo, 76, 244
   postfix (reverse Polish), 187
   prefix (Polish), 187
array, 41, 79
ASCII code, 246
associative law, 35

binary
   code, 248
   number, 73
   positional tree, 161, 178, 183
Boolean algebra, 53
Boolean quantity, 47

cardinality, 28
Cartesian graph, 196
Cartesian product, 39, 194, 208, 220
cats, counting ability of, 71
check digit, 244
circuit, 110
code
   ASCII, 246
   binary, 248
   error, correcting, 246
   fixed length, 242
   Hamming, 262
   maximum rate of, 260
   minimum distance of, 250, 258
   repetition, 249
   self-checking, 243, 245
codes, 241
codeword
   distance between, 248
   length of, 259
   nearest neighbour of, 249
codewords
   interleaving of, 262
combinations, 227
cambinatorics, 20
commutative operations, 31
complement (of a set), 32
composition of relations, 203
computer graphics, 89
connectedness, 107
congruent, 76
counting, 219

decimal numbers, 72
decoding, 243, 250
degree, of graph vertex, 104
De Morgan's laws, 53
digraph, *see* graph, directed
Dijkstra's algorithm, 141
distribution, 227, 228
distributive law, 34

edge, 13, 96
   directed, 98, 124
   pair, 102
   weight of, 139
element
   of a matrix, 80
   of a set, 27
empty set, 28
encoding, 243, 250
enumeration, 219
erasure, 262
error correcting code, 246
error correction, 247
   detection, 247
   syndrome, 255
   word, 247, 254
errors, 241
   burst, 261
   model of, 247 254
equivalence relation, 200
event, 231

Index

events
    exhaustive, 230
    mutually exclusive, 230

factorial, 77, 226
field, 206
file, 206
finite state machine, 10
flag, 22
float, 149
floating point arithmetic, 78
function, 194, 211
    diagram of, 212

generator matrix, 252
graph, 13, 96
    bipartite, 110, 198, 204, 220
    Cartesian, 39, 196
    colouring, 114
    complete, 68, 116, 219
    connected, 107
    directed (digraph), 98, 124, 196
    edge-weighted, 139
    planar, 115
    precedence, 136
    subdivision of, 120
    vertex-weighted, 147
graphics
    computer, 89
    window, 39
graphs, isomorphic, 105
greedy algorithm, 170

Hamming code, 262
head (of list), 43
head clog, 261

identity, 35
implication, 55
indegree, 125
integers, 30
inverse, 35
ISBN (International Standard Book Number), 244
isomorphism, 105, 228
iteration, 20

key, 210
Kruskal's algorithm, 172, 177

list, 42, 144
    append function on, 43
    head of, 43
    linked, 44, 128
    tail of, 43
logical operators, 49
loop, 20, 125

magnification, 90
matrix, 79
    addition, 82
    diagonal, 81
    dimension of, 80
    element, 80
    generator, 252
    leading diagonal of, 80
    multiplication, 83–86
    parity check, 245
    partitioned, 252
    scalar, 88
    square, 80
    symmetric, 81
    transpose of, 81
matrices
    compatible, 84
menu-type problems, 220
minimum distance, 250, 258
modelling, 10
modulo arithmetic, 76, 244
modulus, 202
multiplication principle, 220

natural numbers, 30
necessary and sufficient conditions, 57
network, 147
    precedence, 136
node, 65
n-tuple, 40, 222
numbers
    natural, 30
    rational, 30
    real, 30
number systems
    binary, 73
    decimal, 72
    hexadecimal, 74
    octal, 74

'or' rule, 234
ordered
    pair, 38, 194
    tree, 156
    triple, 205, 220
outcomes, 230
outdegree, 125
overflow, 78

parity, 245
    bits, 248
    check matrix, 245
parity check matrix, 245
path, 110
    shortest, 140
    longest, 145
permutations, 226
pigeonholes, 227
pointer, 44
    forward, 45
Polish arithmetic, 187
post-conditions, 216
power set, 37
pre-conditions, 215
predicate, 48
predicate form (of set), 28

Index

Prim's algorithm, 173
probability, 229, 231
   addition law of, 234
   characteristics of, 223
   multiplication law of, 235
program
   as a function, 213
proof
   by contradiction, 61, 177
   by induction, 64
   constructive, 59
   direct, 60
   enumerative, 64

query language, 207
queue, 44

random number, 223
rational numbers, 30
real numbers, 30
record, 44, 206
reflection, 90
relation, 194
   antisymmetric, 200
   asymmetric, 200
   between two sets, 195
   diagrams of, 196–198
   equivalence, 200
   on a set, 194
   properties of, 198–203
   reflexive, 199
   symmetric, 196, 199
   transitive, 199
relational database, 198, 205
relations
   composition of, 203
repetition, 16, 20
repetition code, 249
rotation, 91

sample space, 231
scalar, 82
selection, 16, 17
   not ordered, 224, 226
   ordered, 224, 226
   with repetition, 224
   without repetition, 224, 226
sequence, 16
set, 25
   cardinality of, 28
   complement of, 32
   element, 27
   empty, 28
   intersection, 31, 235
   power set of, 37
   predicate form of, 28
   union, 31, 222, 234
   universal, 29
sets
   disjoint, 31
   difference of, 31
shortest path algorithm, 144
spanning tree, 153, 169
software engineering, 216
statement, 47
subdivision, 120
subgraph, 109, 118
subset, 29, 231
   proper, 29
subtraction principle, 223
sub-tree, 160
syndrome, 255

tail (of list), 43
tautology, 48
tree, 39, 65, 152
   alternative representations of, 156
   binary positional, 161, 178, 183
   ordered, 156
   search methods, 179
   sorting by, 190
   spanning, 153, 169
   unrooted, 153, 169
truncation, 78
truth table, 50
tuple, *see n*-tuple

underflow, 78
universal set, 29

Venn diagram, 29
vertex, 13, 96
   indegree of, 125
   outdegree of, 125
vertex degree, 104
vertex weight, 147

weight, 13

Zeller's rule, 214

Mathematics and its Applications

Series Editor: G. M. BELL, Professor of Mathematics, King's College London, University of London

Gardiner, C.F.	Algebraic Structures
Gasson, P.C.	Geometry of Spatial Forms
Goodbody, A.M.	Cartesian Tensors
Goult, R.J.	Applied Linear Algebra
Graham, A.	Kronecker Products and Matrix Calculus: with Applications
Graham, A.	Matrix Theory and Applications for Engineers and Mathematicians
Graham, A.	Nonnegative Matrices and Applicable Topics in Linear Algebra
Griffel, D.H.	Applied Functional Analysis
Griffel, D.H.	Linear Algebra and its Applications: Vol. 1, A First Course; Vol. 2, More Advanced
Guest, P. B.	The Laplace Transform and Applications
Hanyga, A.	Mathematical Theory of Non-linear Elasticity
Harris, D.J.	Mathematics for Business, Management and Economics
Hart, D. & Croft, A.	Modelling with Projectiles
Hoskins, R.F.	Generalised Functions
Hoskins, R.F.	Standard and Nonstandard Analysis
Hunter, S.C.	Mechanics of Continuous Media, 2nd (Revised) Edition
Huntley, I. & Johnson, R.M.	Linear and Nonlinear Differential Equations
Irons, B. M. & Shrive, N. G.	Numerical Methods in Engineering and Applied Science
Ivanov, L. L.	Algebraic Recursion Theory
Johnson, R.M.	Theory and Applications of Linear Differential and Difference Equations
Johnson, R.M.	Calculus: Theory and Applications in Technology and the Physical and Life Sciences
Jones, R.H. & Steele, N.C.	Mathematics in Communication Theory
Jordan, D.	Geometric Topology
Kelly, J.C.	Abstract Algebra
Kim, K.H. & Roush, F.W.	Applied Abstract Algebra
Kim, K.H. & Roush, F.W.	Team Theory
Kosinski, W.	Field Singularities and Wave Analysis in Continuum Mechanics
Krishnamurthy, V.	Combinatorics: Theory and Applications
Lindfield, G. & Penny, J.E.T.	Microcomputers in Numerical Analysis
Livesley, K.	Mathematical Methods for Engineers
Lord, E.A. & Wilson, C.B.	The Mathematical Description of Shape and Form
Malik, M., Riznichenko, G.Y. & Rubin, A.B.	Biological Electron Transport Processes and their Computer Simulation
Massey, B.S.	Measures in Science and Engineering
Meek, B.L. & Fairthorne, S.	Using Computers
Menell, A. & Bazin, M.	Mathematics for the Biosciences
Mikolas, M.	Real Functions and Orthogonal Series
Moore, R.	Computational Functional Analysis
Moshier, S.L.B.	Methods and Programs for Mathematical Functions
Murphy, J.A., Ridout, D. & McShane, B.	Numerical Analysis, Algorithms and Computation
Nonweiler, T.R.F.	Computational Mathematics: An Introduction to Numerical Approximation
Norcliffe, A. & Slater, G.	Mathematics of Software Construction
Ogden, R.W.	Non-linear Elastic Deformations
Oldknow, A.	Microcomputers in Geometry
Oldknow, A. & Smith, D.	Learning Mathematics with Micros
O'Neill, M.E. & Chorlton, F.	Ideal and Incompressible Fluid Dynamics
O'Neill, M.E. & Chorlton, F.	Viscous and Compressible Fluid Dynamics
Page, S. G.	Mathematics: A Second Start
Prior, D. & Moscardini, A.O.	Model Formulation Analysis
Rankin, R.A.	Modular Forms
Scorer, R.S.	Environmental Aerodynamics
Shivamoggi, B.K.	Stability of Parallel Gas Flows
Smith, D.K.	Network Optimisation Practice: A Computational Guide
Srivastava, H.M. & Manocha, L.	A Treatise on Generating Functions
Stirling, D.S.G.	Mathematical Analysis
Sweet, M.V.	Algebra, Geometry and Trigonometry in Science, Engineering and Mathematics
Temperley, H.N.V.	Graph Theory and Applications
Temperley, H.N.V.	Liquids and Their Properties
Thom, R.	Mathematical Models of Morphogenesis
Toth, G.	Harmonic and Minimal Maps and Applications in Geometry and Physics
Townend, M. S.	Mathematics in Sport
Townend, M.S. & Pountney, D.C.	Computer-aided Engineering Mathematics
Trinajstic, N.	Mathematical and Computational Concepts in Chemistry
Twizell, E.H.	Computational Methods for Partial Differential Equations
Twizell, E.H.	Numerical Methods, with Applications in the Biomedical Sciences
Vince, A. and Morris, C.	Discrete Mathematics for Computing
Walton, K., Marshall, J., Gorecki, H. & Korytowski, A.	Control Theory for Time Delay Systems
Warren, M.D.	Flow Modelling in Industrial Processes
Wheeler, R.F.	Rethinking Mathematical Concepts
Willmore, T.J.	Total Curvature in Riemannian Geometry
Willmore, T.J. & Hitchin, N.	Global Riemannian Geometry

Statistics, Operational Research and Computational Mathematics
Editor: B. W. CONOLLY, Emeritus Professor of Mathematics (Operational Research), Queen Mary College, University of London

Abaffy, J. & Spedicato, E.	ABS Projection Algorithms: Mathematical Techniques for Linear and Nonlinear Equations
Beaumont, G.P.	Introductory Applied Probability
Beaumont, G.P.	Probability and Random Variables
Conolly, B.W.	Techniques in Operational Research: Vol. 1, Queueing Systems
Conolly, B.W.	Techniques in Operational Research: Vol. 2, Models, Search, Randomization
Conolly, B.W.	Lecture Notes in Queueing Systems
Conolly, B.W. & Pierce, J.G.	Information Mechanics: Transformation of Information in Management, Command, Control and Communication
French, S.	Sequencing and Scheduling: Mathematics of the Job Shop
French, S.	Decision Theory: An Introduction to the Mathematics of Rationality
Griffiths, P. & Hill, I.D.	Applied Statistics Algorithms
Hartley, R.	Linear and Non-linear Programming
Jolliffe, F.R.	Survey Design and Analysis
Jones, A.J.	Game Theory
Kapadia, R. & Andersson, G.	Statistics Explained: Basic Concepts and Methods
Lootsma, F.	Operational Research in Long Term Planning
Moscardini, A.O. & Robson, E.H.	Mathematical Modelling for Information Technology
Moshier, S.L.B.	Mathematical Functions for Computers
Oliveira-Pinto, F.	Simulation Concepts in Mathematical Modelling
Ratschek, J. & Rokne, J.	New Computer Methods for Global Optimization
Schendel, U.	Introduction to Numerical Methods for Parallel Computers
Schendel, U.	Sparse Matrices
Schmi, N.S.	Large Order Structural Eigenanalysis Techniques: Algorithms for Finite Element Systems
Späth, H.	Mathematical Software for Linear Regression
Stoodley, K.D.C.	Applied and Computational Statistics: A First Course
Stoodley, K.D.C., Lewis, T. & Stainton, C.L.S.	Applied Statistical Techniques
Thomas, L.C.	Games, Theory and Applications
Whitehead, J.R.	The Design and Analysis of Sequential Clinical Trials

STAFFORD LIBRARY
COLUMBIA COLLEGE
1001 ROGERS STREET
COLUMBIA, MO 65216